Memoirs of a

Mobile

Diabetic

By Britta McGeorge

Our mission is to efficiently provide the world's finest, most comprehensive book publishing service, enabling every author to experience success. To find out how to publish your book, your way, and have it available worldwide, visit us online at www.trafford.com

Trafford rev. 4/15/2010

Trafford PUBLISHING® www.trafford.com

North America & international
toll-free: 1 888 232 4444 (USA & Canada)
phone: 250 383 6864 ♦ fax: 812 355 4082

Do you recognize any of the symptoms listed below?

Blurred vision

Dizziness or light-headedness

Fatigue

Head-ache

Laughing

Nausea

Nervousness

Slurred speech

Sweating profusely

Weakness

Yawning

[If you are a diabetic, you will.]

Matter- of- fact Verse

Diabetes is no picnic
It's serious, without a doubt
Learning how to manage it
Is what it's all about

With frequent testing, eatery etiquette,
and the proper medication
It can be controlled with lots of care
And visits to your physician.

Listen to your doctor
And implement suggested treatments
That can reveal other methods
More suitable for your daily regiments

Giving up is not an option
Of this, you can be ensured
That there are others just like you
Who have persevered and endured

Dedication

I want to dedicate this book to those who have helped me during my most critical times with diabetes. Dr. Weinburg was my diagnostic doctor and I dedicate my life to him. The people who have supported me the most frequently were my immediate family and close friends. Especially my dad, mom, sister, younger brother, and my friends, Natalie and Eileen. My support system is essential and they are all included in my family tree.

Acknowledgements

I wish to acknowledge my family for their unwavering love and support, especially during critical times in my life. My immediate family includes: my older and younger brothers, my sister, and my mom and dad. My mom married young and was married to one man for her entire life until he passed away in 1996. I would not be here today if it weren't for them. My parents are Wayne and Eleanor McGeorge. My name is Britta. I am the second eldest child of the other three siblings mentioned. There were people who helped me when I was unable to help myself. If they were not good and honest people, things would have turned out differently for me. I want to thank these individuals also and hope that they will remember and become aware of my sentiments. I don't know any of them by name but I am sure that they will remember me. Once upon a time, I stopped them at a traffic light in Virginia Beach, VA. Some of the names mentioned may not be correct but the events being shared are unmistakable.

Need – to – know Acronyms and Words

Blood glucose or Blood sugar- Simple sugar that
is the major energy source in the body.

BG- Blood Glucose

DMV- Department of Motor Vehicles

Diabetes- There are two types of diabetes. Diabetes Mellitus
is a complex and chronic disorder of metabolism due either
to partial or total lack of insulin secretion by the pancreas.

Diabetic- an individual who has diabetes.

EMT- Emergency Medical Technician

Glucometer – Glucose Meter

Ketoacidosis- Critically high blood glucose
that can render the body comatose.

It is sometimes called Diabetic Coma or Diabetic Ketoacidosis.

Hyperglycemia- Excessively high blood glucose
which can lead to DKA; coma.

Hypoglycemia- Lower than normal blood glucose
level which can lead to Diabetic Shock.

Mg/dL- Milligrams of sugar per deciliter of a blood drop,
used when doing blood glucose testing.

NCE- National Certification Exam

PMS- Pre-menstrual Syndrome

Contents

Introduction

After a few years of deliberation, I have decided what I want to write about; My life with diabetes. For non-diabetics, my book will not interest you as much unless you know someone with it. For diabetics, you will relate to events being described and recognize things that you've heard from your doctor that were difficult to believe. However, this is my chronicle and it did happen to me. It involves situations where my lifestyle was affected, directly or indirectly, by my diabetes. However, I would like to give you some information about myself and my family. You can better understand what it was like growing up with diabetes through information about my environment. Medical technology was different then as well. My family and myself were the predominant people who dealt with my diabetes except when I visited my doctor. I've dealt with the daily regiments of my diabetes completely on my own since I was eleven and one-half years old. It has been more difficult to manage financially and I've had some help with this.

I was born at the Halifax Community Hospital in South Boston, VA at 11:38AM on Monday, February 2, 1959. At least, that's what it says on my Certificate of Birth. I weighed in at six pounds and eleven ounces. I was a good size baby for a girl, in that day and time. My father, mother, and older brother were living in Blue Lake, WV at the time of my birth. My mom liked the doctor that delivered my older brother and continued to see him even after my dad was relocated from Virginia to West Virginia. It was February and the weather was cold in South Boston and Blue Lake. West Virginia generally gets more snow than Virginia.

My mom remembers a day in Blue Lake during the month of March when there was a lot of snow. There was so much snow that the doors to our house were frozen shut for three days. My dad had resorted to climbing out of the windows to get to work. He was a lineman and had to work on those electrical lines to restore electricity to homes and businesses. I was three years- old when we moved away from West Virginia, according to my mom. Once we left West Virginia, we never returned to live. We did return twice to take a summer vacation during my childhood.

I was born with brown eyes and brown hair. My skin is white and turns coppertone brown in the summer from the sun. Strawberry blonde was the color of my hair from summertime swimming and other outdoor adventures. As a very active diabetic in my childhood, my summers were frequently spent outside, riding my bike and playing in the creek with my brother. I got yelled at recurrently for coming in the house with wet feet. I started taking my shoes off at the back door. The creek that ran in front of our house was small and made an L-shape around our house. Our property was kind of triangular in shape.

We lived at the bottom of a cul-de-sac and the front of our house faced the street. There weren't many cars that traveled through our cul-de-sac. I rode my bike and played there often. I used to walk to the mall which was a mile from my house whenever I was allowed to go. I loved to shop because I was so small that I could wear anything. However, I did a lot of window shopping because we didn't have the funds to buy all the pretty things I saw in the stores. During my pre-teen years, everything I saw was pretty.

Television advertised so many beautiful clothes, jewelry, and shoes that I was always admiring something. Sometimes, I couldn't find the exact item that I wanted but I usually found something similar that I liked just as well. When I told my mom about what I saw and that I couldn't afford it, sometimes I would find it under the tree at Christmas or in a gift box for my birthday. Not always but I was always happy with whatever they gave me. Keep in mind that I was still your typical teen-age girl. If it didn't fit or I didn't like the color, I took it back. The public library was located before the mall and I often stopped by there and never made it to the mall. I loved to read and enjoyed having so many books to choose from in the library. Going to the library is still a favorite pastime of mine. Being able to use their computers is a plus too. I went to school during the weekdays and went to the mall or library on weekends, when I wasn't at my grandparents with my dad.

On weekends, I often went with my dad to his parents trailer out in the country to visit with them. My grandparents lived on a huge farm. My older brother came sometimes and that was when I got into the most trouble. Their farm came complete with a large garden, a barn with a fenced in pasture with grass for their horse, and they had some pigs that stayed in a pen near a barn-like structure in the back of their trailer. My grandparents lived in a double-wide trailer in Clarksville VA and they had stray dogs, tractors, and an old beat up truck. They also had their own gas pump in the back yard.

My older brother didn't visit often because he said there was nothing to do on the farm. I had to go because there was no one else to administer my injections until I was eleven and one-half years -old. Therefore, I usually played by myself on Friday and Saturday and went to church with my friends on Sundays. I helped with the garden and that involved most of the morning and early afternoon hours on Fridays and Saturdays. By lunchtime, it was too hot for us to continue doing anything in the garden. After we picked vegetables from the garden, we had to prepare them for dinnertime.

This particular garden had corn, tomatoes, green beans, cantaloupe, and numerous other fruits and vegetables. It consumed about two acres or more of land so there was always plenty to do after we picked fruits and vegetables from the garden. I also saw snakes in that garden frequently and became afraid of them. They were so long, black, and scaly looking that it was creepy. My older brother used to like to tease me about being afraid of those snakes. If it was a snake, I didn't care what kind it was. I didn't like it. When I got to middle school, I had a friend who had a snake as a pet. I never went anywhere near it. That garden relieved me of any snake fettish that I might have harbored.

My grandfather had grown tobacco too and had barns located on his land where the tobacco leaves could dry. I used to assist them in bunching up those leaves and tying them together to hang over a pole in one of the tobacco barns. It took those leaves almost a week to dry and curing tobacco was time-consuming. After I finished with the garden and tobacco by the afternoon on Friday and Saturday, I would spend the evenings with my dad teaching me how to play the card game Solitaire. We played double solitaire too. Most of the card games I learned are much easier to play on a computer now. You don't have to shuffle and deal the cards. The computer does it for you.

There wasn't much else to do in the country at night. Sometimes, we didn't even have television. If we did have reception, we were always watching the news. I was a kid and didn't have much interest in the news but would watch it with my dad and grandparents anyway. I have the card game Free Cell on my computer now and can play it continuously at home. When I stop playing Free cell, I will either cut on the television or pick up a book and read. Things are very quiet around here, especially after dinner. We don't live in the country now but I really enjoyed that type of life. It was quiet and peaceful, kind of like when you go fishing and leave your cell phone at home. My country side comes alive when city life gets too busy.

This book is autobiographical and focuses on my life with diabetes. Some of the events mentioned are being told from a medical professional's conversation with me after awakening from a diabetic episode. My life was changed forever after I was diagnosed with this disease. Diabetes is a disease and is life-threatening IF it gets out of control, type 1 in particular. A healthy diet aids in keeping diabetes under control. It is very difficult to be a diabetic in our fast-paced world today. It takes time to cook and prepare foods according to how my doctor and dietician recommended.

Cooking food in a microwave wasn't recommended as the best way of preparing my food. My mom was great at cooking and implemented what foods they wanted me to eat into my diet. There were so many things that she had to pay attention to when making meals for me. Concentrating on meals with a low-sugar content was the priority. Grocery shopping became a real job and that hasn't changed. Reading the nutritional labels takes time, in order to make a good choice. However, price factors into this choice and I make the best choice possible using what I know about food as my guideline. People are becoming more health conscious when it comes to choosing the foods that they eat.

There were things that happened before I was diagnosed with diabetes but I can't remember many of them. The diabetes was such a tragedy in my life that any pleasant memories before that time were quickly forgotten. The diabetes involved meticulous care and attention on my part and everyone else around me and dominated my childhood lifestyle, dreams, and memories. My lifestyle as a child seems serene but dealing with the diabetes was relentless. The two environments balanced each other well. My family has always been there when I needed them. As I've aged, my younger sister and brother have helped me a few times with my diabetic episodes. I am the only family member with a chronic medical condition. I always say that I am not perfect in the way that I treat my diabetes but I am still here to talk about it.

Chapter 1: Diagnosed

I don't remember much of anything about my life before my diagnosis in July of 1968. It was so catastrophic that it paled all memories, good and bad, before that particular time in my life. I never suspected how complicated my life was about to become. I found that hyperglycemia and hypoglycemia caused by diabetes could distort my thinking processes and affect my attention span. Hypoglycemia occurs more frequently and requires my immediate attention. I found that these two aspects of diabetes were going to complicate my life more than anyone could ever imagine. Even myself.

It started one hot summer day on Monday, July 15, 1968. I had been laying on our tan, leather den sofa for a few days with a slight fever and chills. I was constantly thirsty and going to the bathroom and, eventually, started vomiting. The vomiting started in the morning and got worse by lunchtime. I had started vomiting uncontrollably and was having dry heaves. My mom and dad thought that I had the flu. On that Monday, my mom called my dad at work. She spoke to his secretary who said that he was in a meeting. She interrupted his meeting to let my mother talk to him. He worked full time for an electric company and had been there for thirteen years. His job was very demanding and he had time that he could use if an emergency ever arose. This was an emergency situation.

She told him that my fever was worse and that she couldn't keep me awake. She was having trouble getting me to respond to her when she said my name and I had been vomiting. She told my dad that he needed to take me to the hospital. Mom couldn't drive at that time and my older brother was eleven years old and not old enough to drive yet. My dad was our sole source of transportation.

My dad got home about thirty minutes after my mom called him. It was around noon and we would have to struggle with some lunchtime traffic on the way there. It took my dad five minutes to wake me up and I groggily stayed awake long enough to get my shoes on. My dad helped me with my shoes while my mom got a few clothes together for me. I had been sleeping on our den sofa in the same clothes for the last two days. My favorite t-shirt and shorts were what I was wearing when dad and I left the house for the hospital that day. My dad asked my mom to call the hospital and let them know that we were coming.

I remember that I was able to walk from our den to the front door without too much distress. But I was faint and couldn't turn the doorknob to get onto our front porch. My muscles had gotten feeble from laying on the sofa sleeping for almost a week. My dad turned the knob and assisted me in walking onto the porch landing. My knees were weak and I was very unsteady on the steps. I was dizzy from the fever too. By the time I got to the third step, I vomited over the railing and my dad helped me to sit down. There were only five steps, including the landing, but my body was feeble.

I was miserable. I was going in and out of consciousness and he asked me if I felt well enough to walk to the car. It was only about ten feet away from us. I said that I would be all right and when I tried to walk, I lost my balance again. He put his left arm around my waist and held my right hand in his right hand while we walked to the car and he sat me in the front seat next to him. My mom couldn't come with us because she needed to be there when my brother got home from school and take care my younger sister. So, my dad backed the car out of our driveway and headed straight for the hospital. We were there in twenty minutes. I was completely unconscious by then and he had to carry me into the hospital. A gurney was at the door waiting for us as we entered the hospital.

The doctor that treated me knew that I had diabetes when he saw us coming into the hospital entranceway. Once my dad brought me through the hospital doors, I was placed on the gurney.   I had lapsed into diabetic ketoacidosis and was unconscious and in a diabetic coma. I am sure that my doctor performed tests to confirm his diagnosis of diabetes. The tests they used in 1968 weren't as advanced as those of today and took longer to process. As soon as he was certain, he dispensed short-acting insulin to reduce my glucose level as quickly as possible. However, had my blood sugar level continued to rise, I would have been dead approximately an hour later. Beginning on that day, I became insulin - dependent for the rest of my life. I am not exactly sure how long it took him to confirm his diagnosis but he saved my life with his findings. When I first awoke from my diabetic coma on my fifth day in the hospital, my mouth was still dry and I was exhausted. But my stomach wasn't nauseous anymore. Just sore. I was beginning to feel like my old self again. Lunchtime was approaching fast and I was starved.

I had a hospital lunch of spaghetti with a small side salad and unsweetened tea. I asked the person who brought my lunch to me why I had to drink unsweetened tea? Normally, I had my tea sweetened with sugar at home. She said that it was indicated on my menu by the doctor and I would need to ask him. I asked her what her name was and she said that it was Laura. She also said that she and I would be working together until I went home. She said that the doctor would be in around my dinnertime and I could ask him any questions at that time. His name was Doctor Weinburg. Since I couldn't have sugar to sweeten my tea, I asked her what I could have? She told me that I would need to use sugar substitutes. Back then, I called it, "The drippy stuff." (At dinnertime, I was introduced to Saccharin.) I was so hungry that I inhaled the spaghetti. I don't remember what else I had because my mind kept wandering back to the sweetener comment. I couldn't imagine what dinner would bring and could hardly wait to talk to my doctor. I knew my dad would be by to see me after he got off from work at dinnertime and I was looking forward to seeing him too.

I was in the hospital for exactly two weeks and five days. The next two weeks were spent educating me and my family about how to deal with my diabetes. I spent those next two weeks in the hospital walking up and down the hallways three times a day. They gave me a twenty page book about giving shots and I believed that the shots would stop, once I got home. I didn't pay much attention to that book, until they further explained that I would need to take shots every day for the rest of my life. I was in denial and didn't want to believe this. I couldn't understand what was happening to me and why. This time in my life started as a tragedy. The more I learned about the disease and dealt with it every day, the more I realized that my life was changing in a very big way. As time progressed, it became a part of my lifestyle and became much more manageable.

I was diagnosed as a type 1 diabetic with diabetes mellitus. Diabetes mellitus is a complex and chronic disorder of metabolism due either to a partial or total lack of insulin secretion by the pancreas. This meant that I would need to take a shot every day for the rest of my life. Also, this would require me to follow a strict diet so that I wouldn't get sick from eating too much sugar. If I did eat too much sugar, I would need a shot of insulin to regulate my blood glucose level. Exercise helps the body to burn sugar and metabolize food without utilizing any insulin. This was all that I understood after two weeks and five days in the hospital. It was obvious how serious this illness was after what had happened to me. The basics were comprehensible to me but this was just the beginning. My diagnosis was conclusive and I had a long curvy road ahead of me.

Chapter 2: 1968 Diabetic Education

Education about diabetes was extensive. There were many things that I needed to know about caring for myself. They overwhelmed me with information but I was a quick learner at age nine and one-half. I spoke with various nurses, dieticians, and my doctor about my specific needs. There were two nurses, Laura and Helen, that I became friends with quickly and we worked together to help me understand all of this. Laura was a dietician and Helen was a physical therapist. Every day of that two weeks got a little bit harder and easier at the same time. All I knew was that I wanted to go home so that my life could go back to what it was. Unfortunately, this wasn't going to happen for me and I didn't realize this until I got home.

After two weeks and five days of education, I learned that the three most important things I needed to focus on were: diet, insulin, and exercise. My mom and dad were given an abundance of education about diabetes while I was comatose. My dad knew that he would need to give me injections because my mom was unable to do this. My mom took on the responsibility of cooking for me. I had to remember to keep diet, insulin, and exercise balanced on a daily basis to prevent further complications in my life. I realized as my education continued that this was much easier said than done. My mom was unable to learn how to give me a shot and neither was I. I don't know why this was so difficult for me but I discovered that it was going to take a lot of courage for me to accomplish this task successfully. Therefore, I spent most of my time concentrating on diet and exercise.

I remember that one day, I went to school and my mom had packed a pimento cheese sandwich for my lunch. I was in the fourth grade at that time and having some issues with my new dietary restrictions. I never liked pimento cheese but it was on my list of things that I could eat, I think. Mom and I talked about this before I went to school that day. I thought that I'd be okay eating it just that once and couldn't make myself eat it when lunchtime came. Then I got scared because I knew I wouldn't be able to eat again until dinner. The doctors and dieticians had been adamant about ensuring that I ate breakfast, lunch, dinner, and a snack before bed every day. I knew that I had to eat something before dinner if I didn't eat any lunch. I told my teacher about it and she called my parents house and spoke to my mom. She said that they were having fried chicken for lunch that day and asked my mom if this was ok for me to eat. She said YES and I was elated. The doctors were clear about the kinds of foods that I couldn't have. What I could have was more of a mystery to my dad, mom, and I.

Everything turned out all right that time, even though I wasn't supposed to eat fried foods. I was finding that it was ok to eat ill-advised foods but in minimal quantities. That day was only a preview of the things to come in my life. I found out later how my condition was affecting people's thoughts, feelings, and behaviors around me. Therefore, as my life changed, so did everyone else's. I found that I made choices to do things that revolved around the demands of my diabetes. For example, if I was given a choice, I would choose the activity closest to the cafeteria or snack machine while on school premises. Food was a priority in hypoglycemic situations and I did my best to be prepared. On weekends, I would decline to go to a certain restaurant with a friend because I knew their restroom door locks were all broken. I would choose another restaurant that everyone liked and I knew had better restrooms.

Dinner required me to take a pre-dinner injection. I didn't want anyone who didn't know me to see me taking an injection. I didn't want to use a restroom stall without a door lock on it regardless of whether I was giving an injection or not. My friends who knew me would not have been alarmed but a stranger wouldn't know this. OR I would eat only vegetables because I knew that they didn't require any insulin for metabolism and consumption. This meant that I wouldn't need to take any insulin right then.

I often tried to make sure my vote was cast for my favorite restaurant and there was frequently a medical reason for that choice subconsciously. It was easier to make food choices than to try to change the physical accommodations, like the restroom door locks. Food Courts were just becoming popular in that day and time and I enjoyed hanging out in the malls with my friends. My all time favorite weekend activity was roller skating. Their concession stand continuously had something that I could eat. I was so active as a child and teen-ager that I burned numerous calories through physical activities. In school, I loved to play volleyball and usually packed something for a snack. I found that healthy snacks were and are a great way to handle these situations.

Eventually, I learned how to enjoy my favorite activities without having to change my physical accommodations. Ie: going to a bowling alley that included an arcade. I found ways to deal with my diabetes that didn't exclude me from the things I liked to do. But I also had to remember that my diabetic regiments couldn't be ignored. There would be consequences to face, otherwise. Finding that delicate balance between diet, insulin, and exercise was the challenge that confronted me every day.

Spontaneity doesn't concur with a diabetic lifestyle. Mealtimes especially necessitate pre-planning. If I am going to consume a meal high in carbohydrates, then I need an adequate amount of insulin to consume it. If insulin is taken now, I must eat within ten minutes thereafter. Therefore, I need to know about meals ahead of time. Being prepared for mealtimes is an art form to a diabetic. A diabetic should never be caught without a medic alert device. This prepares them for that unexpected hypoglycemia. I have several of the stainless steel chain necklaces that I can wear under my clothes, as opposed to the bracelets. However, my necklaces are not usually recognized until after an incident. The bracelets are used more often. Spontaneous events happen and are easier to handle once you understand them.

Words of Wisdom: Diabetic Education has come a long way since 1968. There are different medical devices and medical resources available to make life less regimented and more practical. Unfortunately, a lot of what you learn has to come from your own experiences with diabetes. Now, in 2009, I am on an insulin pump and will continue to use it as long as possible. Taking daily injections is difficult to maintain in any working environment and in daily life activities. I know this because I was injection dependent for thirty years. AS ALWAYS, if you are a type 1 diabetic and want to utilize an insulin pump, you should consult with your doctor.

Diabetic Education Continues

Chapter 3: Food

Food is a part of our every day life. It is necessary to fuel our bodies and sustain our life. It is the controlling factor in the dietary part of the diabetic equation: Diet, Insulin, and Exercise. The amount of food and what type of food I eat controls how high or low my blood sugar becomes, combined with how much insulin I take for that meal or snack. Keeping normal blood glucose levels is my ultimate goal. Every time there is a celebration of some kind, food is included in that celebration. I love food. Being aware of the different kinds of foods and their contents is how I determine how much insulin will be needed to consume it. It sounds complicated but really isn't.

Starches are usually the highest in calories, carbohydrates, and sugars, depending on if they are simple or complex. Nutritional labels help to simplify this for me. Many foods don't come with nutritional labels, like your fruits and vegetables in the produce department. This is because they contain the good, or complex, carbohydrates and use a lesser amount of insulin for consumption than the bad, or simple, carbohydrates unless they are cooked. Added oils and sugars increase their caloric value.

We use food to reward children when they have been good. Choosing the right foods to eat can be difficult in certain situations but it can be done. For example, if you are in a pizza parlor, you can choose to eat a submarine sandwich instead of pizza. A submarine sandwich has lettuce, tomatoes, and onions which is much healthier for you. Many pizzerias have tossed salad food items on their salad bars as an alternative food choice.

Peer pressure makes choosing the right foods harder because almost everyone except you will be eating pizza. Pizza is cheaper for the pizzeria to produce and is sold in larger quantities than the salad bar items. The vendor passes these costs along to the consumer which was me, back in 1968. Like any health-conscious diabetic should know, eating good food doesn't have to be expensive. It just takes you a little longer to make a good choice.

Food is used to commemorate many things. New jobs and expected stressful days at work are perfect examples of this. Food keeps us satisfied and reduces our stress levels. If we are not hungry, then we are less likely to take our grievances out on the people around us or even a customer. Hence, the term comfort food means that food can relax us and make us more comfortable and feel less stressed. A comfort food item for me is coffee. For someone else, it may be a candy bar or something entirely different. As a child, food was used for every holiday on the calendar and many that weren't. The ones that weren't included: birthdays, weddings, church socials, after-school events, and etc.

I was excluded from many of these events during the first year after I was diagnosed with diabetes. Even at school because people were afraid of what would happen to me if I ate something that would make me sick. They were afraid that they wouldn't know what to do. When my friends celebrated their birthdays, I was usually excluded from their parties because cake is high in sugar content and was restricted from my diet. As a young child, I thought I had been excluded from their party because they didn't like me. However, I did go to parties hosted by my closest friends. They would have angel food cake or pound cake and diet soda, just for me. They always made me feel special and they will remain in my fondest memories.

I entered the fourth grade after the summer I was diagnosed with diabetes. I finally discovered that my teacher was hesitant to have me in her class because she was afraid that she wouldn't know what to do if something happened to me while in class. Fourth grade was one of my worst school year's ever. I liked my teacher very much, regardless of how she felt about my situation. Her name was Mrs. McGuill and she was always nice to me. And to make matters even more interesting, Mrs. McGuill was promoted to the fifth grade the following year and I was in her class again. My fifth grade year was much better and she seemed more at ease with me in her class than the year before. It was not until I was entering college that my mom told me about how my fourth grade teacher felt. I could empathize with her feelings as I was entering college. I knew how hard it was to understand and be prepared for diabetic episodes, especially during that time.

As I was promoted to sixth grade, I had much better control of my condition. I didn't have as many problems that year and was looking forward to attending middle school the next year. I learned a few tactics that helped me to avoid hypoglycemia and started to keep something sweet with me at all times. This was discouraged during my fourth grade year with diabetes because I liked candy and would eat it whether I needed it or not. During my childhood and teen-hood, I became more adept at explaining things to people to deter their fear of me. I found that they were better able to understand and accept it when something did happen. Everything was easier to deal with as a child than as an adult. This is true for most people that I've talked to.

Today, as an adult, there was a Kentucky Derby party that I was invited to. We all entered $.50 in the cash pot and picked the name of a horse to win. With every $.50 we put in the pot, we were able to choose another horse and improve our chances of winning. There was a lot of food served at this party. We had hamburgers and hot dogs, pasta salad, green beans, cottage cheese dip, deviled eggs, baked beans, strawberry shortcake and chocolate cake. There were some other things on the table and since I didn't know what they were, I chose not to eat any of them. I ate a hot dog on a bun with some of the pasta salad, green beans, and deviled eggs. The pasta salad was a little questionable but it contained mayo, pasta, tomatoes, and some green peas. I drank water with mine because everyone else was drinking 'spiked' punch.

The one thing you have to remember about food is that eating something in moderation is always a good choice. Even bad foods, in moderation, have some good things in them that your body requires to remain healthy. Choosing the right food is always a challenge, no matter where I am. I may not always make the right choice and, sometimes, it is easier to choose something by knowing what NOT to choose. This choice has become simpler and has formed some very good eating habits through the years.

Remember this the next time you have a party and try to include dietary or sugar-free foods for people like me. In my childhood, I was taught that coffee and diet sodas were fine for me to have because they contained NO sugar. Some diabetics may choose some other diet drink for their beverage or even water. I've always avoided any drink containing sugar or excessive calories. If you are unsure if a soda contains sugar or not, check the ingredients. If it contains sugar, sugar will be listed among one of the first three ingredients. If you are still unsure, choose another beverage. Contact your dietician concerning any dietary questions you experience as soon as possible. Nutritional labels help to make any party more enjoyable for me.

Chapter 4: My First Birthday After My Diagnosis (#10)

This was my tenth birthday on February 2, 1969. We celebrated it on the Friday before my birthday on that Sunday because several families had plans to leave town for that weekend. I didn't mind this if everyone who said they were going to come did. Out of the thirty or so friends and families that I asked to come, about ten of the children I truly wanted to come attended. I had wanted to have my birthday party at Bronco's Barbecue since right before I was diagnosed. My mom tried to explain to me that my doctor would not approve of me eating anything considered to be fast-food. She felt that the barbecue would have too much fat in it, according to my doctor. She said that I would not be able to eat the barbecue at Bronco's Barbecue because she didn't know exactly how it was prepared. Therefore, she would need to pack me a lunch with something that I could eat at the party, if I insisted on having it there.

I wanted this as much for me as I did for my friends. It had been a rough seven – eight months for me and I felt that my friends would be less afraid if my parents were there. They would be able to see that I could eat out somewhere and it wouldn't make me sick. I really wanted to have my birthday party at Bronco's Barbecue and my parents seemed reluctant, at first, and finally gave in. I could tell that my mom was still against the idea but it was my birthday and my parents wanted me to enjoy my birthday.

I invited several of my friends from school and a few people that I knew from our church. About ten children and their parents showed up. Back in 1969, when a family was invited to attend such an event, that included the mother, father, and children. Pets remained at home but the family was all there. I had one friend who I wanted to come with her mother but didn't because she didn't have a father and respectfully declined my invitation. I missed her at my party but understood. I knew that everyone I invited wouldn't be able to attend and was glad to see those that did. Since it was a few hours after we had left school, most of the kids were wearing what they had worn to school that day. Some of them had changed clothes but not many.

The fathers were mostly wearing docker-type slacks and a nice collared shirt, like my dad. Most of the mothers were wearing a nice skirt or pair of slacks and a pretty blouse, like my mom. I was wearing my favorite jeans and a pretty powder - blue top with somewhat white sneakers that my mom let me wear. My older brother wore a pair of jeans and my mom made him wear a nice collared shirt. He wanted to wear a t-shirt but mom made him wear a collared shirt over the t-shirt. He liked barbecue and I think he enjoyed it too but it took me a week to get some sort of opinion out of him. He said that he liked it and didn't use any enthusiasm when he said it. I decided I'd remember that for his birthday in June.

This was the type of barbecue place where a waitress roller-skated up to your vehicle and took your order. My dad was driving a tan and white colored Plymouth, vintage- looking. I don't remember that much about it because I wasn't into cars much during that time. It had green buttons that lit up at night on the dashboard. My family loved that car, especially my dad. We were all sitting in our vehicles when the waitress came to our car. My dad ordered a Tab diet soda for me to drink with my lunch. Everyone ordered barbecue except me. There were a few picnic tables surrounding Bronco's Barbecue and some of the kids and parents got out of their cars to sit around the tables and eat.

I ate what my mom packed for me and didn't complain because I promised her that I wouldn't. It didn't take me long to eat my tuna sandwich and I took my drink over to the picnic table and sat with my friend's while they ate and talked. I wanted everyone to have a good time and ease some of the tension created by my sickness. Everyone seemed to enjoy this outing. After we ate, the people sitting at the picnic tables discarded their trash in nearby receptacles and we all remained seated around the tables and talked for awhile longer. This is when we decided to meet at the Sun strip bowl to have some fun bowling. The Sun strip Bowl was located down the street about ten blocks from Bronco's Barbecue. When everyone had finished eating and all trash had been discarded, we piled into our vehicles and headed for Sun strip Bowl. Once we arrived there and started bowling, it reminded me of the times I had been bowling with some of them before.

I admit that I was a little disappointed that I couldn't eat their barbecue but I was happy that everyone enjoyed it. My birthday was on a Friday afternoon and the kids were still talking about it on Monday. I think it had helped to ease my friends and their parent's minds about my situation in dealing with my diabetes. People that I knew started behaving more like they were before I was diagnosed. Some of the kids continued to ignore me, like they did before, because I wasn't in their clique group and wasn't a popular kid. I felt better when sharing information with my friends about my diabetes after this outing. Ultimately, it had turned out to be a very normal and much needed day for my family, friends, and myself.

Chapter 5: My Eleventh Christmas

This particular Christmas event had nothing really to do with my diabetes except maybe subconsciously. It reflects how stubborn I was and how much I believed that I could be rid of the diabetes, one day. I didn't know how it would happen but I had heard stories of others who had surpassed it and hoped to be one of them. This story happened during my eleventh Christmas holiday that I would like to share with you. This was one of my favorite Christmases ever. We had started a tradition when I was age eight where we could open one present the night before Christmas day. We talked about this tradition when I was seven and age eight initiated this tradition. I looked forward to it every year thereafter. But this year was different.

I had developed an interest in mechanical stuff. My older brother had gotten a really neat train set that year, which would have made him thirteen for his birthday. We had played with it together and this is where I think my interest began. Anyway, my mom, dad, brother and myself decided together which present we wanted to open first on Christmas eve. My mom and dad knew what they had bought us for Christmas and mom chose the gift that I had given her and dad chose the gift that my brother had given him that year to open for our Christmas eve tradition. My brother opened his present first and it was from dad. He had gotten my brother a baseball glove from the baseball stadium where they had gone together to see a game. My brother loved that baseball glove He wanted to go outside and play with it but it was too dark for that. My dad was glad to see that he liked it so much.

Then it was my turn to open my present. I had chosen a small, heavy box because I couldn't imagine what it might have in it. My dad told me that I might not want to open that one first because it went with something else that he had gotten me. I was so curious to know what was in that small, heavy box that I opened it first anyway. Well, when I opened that box, it had a black, battery charger in it that you plug in the wall. Dad further explained that there was something in another box that I'd been asking for that went with it. It was funny because I was too curious about that box not to open it. Then when I did, it left me thinking about what he could have possibly gotten me that would go with a battery charger all night.

My dad had gotten a set of nail clippers from my brother that my mom helped him to pick out for dad. It was more like a grooming kit for men. I had given my mom a mug that said World's Greatest Mom on the front of it. My brother got a baseball glove. And I.got a battery charger. My present was the only one that came with another gift that would make it complete. My dad tried to tell me that I should have chosen something else but my curiosity got the better of me. It wasn't a bad choice but it could have been a better one, if I'd listened to him. There were many other presents under the tree for me, mom, dad, and my brother and that special one was sitting there just waiting for me to open it. I was so excited about the prospect of that other present that I only slept maybe four hours that entire night. I was going to have a very special gift to open that would go with my charger in the morning. The night was going to be a long one and seemed never-ending.

When Christmas morning finally arrived, I was the last one downstairs. I was extremely exhausted from my sleepless night but had to know what went with my battery charger. Before I opened it, my dad told me that it was something that I had been suggesting that I wanted for more than a month. I didn't have any idea what it might be because I couldn't think of anything I'd asked for that required a battery charger for usage. I picked up the big box and found that it was a little heavy too but lighter than the battery charger. You'll never guess what was in that box when I opened it. It was a tape recorder.

The year was 1970 and technology wasn't what it is now, in 2009, but I loved that tape recorder. I started taping songs from the radio. I found that voices were the hardest to record, especially when someone was talking too low or quiet to be heard or their words ran together and didn't make sense. After a month, I started to hear a buzzing noise frequently when I played something back. Even some songs recorded from the radio sometimes had buzzing noises in the background. This just goes to prove how far technology has come.

I thoroughly enjoyed that tape recorder, even with it's limited ability to produce good sound quality verses from songs. My main reason for wanting a tape recorder was to tape my favorite songs so that I wouldn't need to buy tapes from the stores. There were so many background noises that the quality wasn't as good as a store bought tape of my favorite songs. I used the tape recorder to tape everything until it broke. I'd had it for almost seven months when this happened.

I remember that I accidentally knocked it off of my dresser and onto the floor one day. It never worked the same again afterward. That was when the buzzing noise became louder. The little door that you opened to insert a new tape into the machine wouldn't close securely anymore. I loved that tape recorder and was hurt when I couldn't get it to work properly. I went to one of my friends to see if he could fix it for me. He tried but it was pointless. I tried to fix it myself several times and finally admitted to mom and dad about what had happened. I didn't want them to find out about it until I had fixed it. Since that never looked like it was going to happen, I had to tell them. That Christmas present was one that I've never forgotten.

My curiosity outweighed my better judgment on that Christmas eve. I knew my dad was trying to tell me something and I couldn't resist my temptation to open that box. I couldn't imagine what was in that box. Celebrating Christmas eve and Christmas day the way we did made the holiday more enjoyable. And my mother's birthday is on December twenty-seventh. A lot of Christmases have passed since then and they have not all turned out as well as those of my childhood. To this day, Christmas remains to be my favorite holiday.

That Christmas also came with an abundance of food. I had a great deal of will power by that time in my life and only ate one brownie in that entire week after Christmas day. I had learned from my doctor and the nurses how bad it was for me and had learned how to eat what was good for me so that I was satisfied. Also, I had to keep my weight under control so I chose foods that wouldn't cause as much weight gain. It required will power to stay away from those plain M&M chocolate candies that I liked so well.

There is always an abundance of food during Christmas and eating smart is a true test of using what you know about food to accomplish the results you desire. I began to learn that there was more to Christmas than just the food and the presents. The people who made all of it possible, my mom and dad and our family, were the real blessings of the holiday. Without them, Christmas would have been empty and meaningless. Family makes holidays more worthwhile. I was very much looking forward to our next Christmas together.

The Unexpected

Chapter 6: Seizures

This is a part of my life that is very hard for me to talk about. I lost so many things due to the seizures. It was hard enough having to deal with the diabetes but now, I had to deal with having seizures too. I had my first seizure when I was a teenager at thirteen. They were grand mal seizures and required immediate attention when they happened. I would thrash around for approximately one minute and then, sleep for almost thirty minutes afterward. They were embarrassing if I was in a public place. It looked like I was having a temper tantrum, or something. Seizures in public were so infrequent, maybe two, that it never really presented that much of a problem.

I was told by my physician that stress and sleeplessness could instigate them. According to him, the diabetes would irritate the seizures and vice-versa. So I tried to get enough sleep and made an even greater attempt to control any possible stressful situations. Controlling stress was difficult because it really cannot be controlled and can happen intermittently all day long.

It's hard to imagine at thirteen what brought on any stress in my every day routines. I started planning for everything weeks in advance, or at least, as soon as I knew anything about what was coming. I made sure that my clothes were laundered regularly so that I always had something decent to wear whenever needed. I took my showers before bedtime. I layed out my clothes for school every night at bedtime. If I was meeting at someone's house the next day, which happened infrequently, I or my mother would call ahead to make sure that the parents knew or were reminded that I was coming. I started carrying more things in my book bag to help organize my days.

My book bag contained: school books, school notebooks, clothes, make-up, candy, and a calendar book to record important dates for me to remember. I was always active in school events such as the book club, volleyball team, and some track events. Anything I could do to prepare for the next day, I did before I went to bed the night before. It helped dramatically but I found that I couldn't prepare for everything. However, these small steps were all important in keeping stress to a minimum. Being careful became habitual for me.

I went through so much from the time I was nine to thirteen. I learned how to deal with the diabetes and now, I had to learn how to deal with the seizures. What was going to happen to me next? I waited for another bomb to drop. But the more I learned about my seizures, the easier they were to control. Learning to control the diabetes and the seizures brought my life to some semblance of normalcy. I found that with the seizures, they occurred most frequently first thing in the morning. I believe it was because I had gone all night without eating and was hypoglycemic as soon as my eyes opened in the morning.

As a child, low blood sugar happened often. I would have a day where it seemed that no matter how much sugar I ate, my blood sugars refused to comply sometimes. I was eating and my body was burning every bit of it. There was a lot of hormonal activity transpiring in my body as a teen-ager. That was another part of my life that was difficult to control. I only had so much control over stress and my hormones. I had started menstruating at age twelve and the seizures started soon thereafter. My life continued to get more complicated as I changed, emotionally and hormonally.

A seizure happened one day approximately eight years after I started having seizures. It was in the middle of the afternoon and when I was on my college campus. I had finished eating lunch shortly before this episode and didn't believe that it could have been due to low blood glucose. After lunch, I had a class at 1:00. I was in the cafeteria downstairs and my class was upstairs in that building. Jerry had been playing chess and borrowed the chess board from Mike who was in my next class. He asked me if I could take it up to him. I agreed to do this and gave it a second-thought afterward. Once I noticed how many books I was carrying and my book bag, it wasn't such a good idea that I tried to transport that chessboard to class.

Jerry had already walked away and I was in a hurry to get to my class. My books were at eye level and with the chessboard placed on top of them, it was really hard to climb the stairs. The elevator was broken and the stairs were my only other option for getting to class. I started feeling funny halfway up those stairs but couldn't leave the chessboard on the stairs. I made it to the top of the steps and someone opened the door for me. I was doing pretty well and still had twenty minutes before class was supposed to start. I turned and started walking toward the hallway that lead to my classroom.

When I was standing in the hallway and only two classrooms before mine, I dropped all of those books and the chessboard when the seizure began. I fell forward and broke my two front teeth. My front teeth had caught the corner of the chessboard and it was wedged between those two teeth. The left one was broken completely in half with the pulp dangling out of the tooth. That is the worst throb and head-ache that I've ever experienced. I laid in the floor very close to the wall for a short while. When the seizure ended and the tremors subsided, I sat up against that wall until the ambulance arrived.

There was a girl in my class, Stephanie, who saw what was happening to me and called an ambulance. She waited until the seizure ended and sat with me for a few minutes. She told me that she had called an ambulance. I shook my head to thank her and motioned that I would remain seated there until they arrived. It hurt too much to talk. I was alone when the Medical technicians arrived and they wanted to take me to the hospital. They helped me up and asked me to get on the gurney for transport to their ambulance downstairs. I tried to tell them that I needed to go to the dentist but couldn't talk. They took me to the hospital first anyway and my dad showed up about thirty minutes after I arrived at the hospital.

I had given them my emergency contact information while in the ambulance. Once my dad signed the hospital discharge papers, we went to my dentist for emergency purposes. My dad had called him to let him know why we were coming. He was prepared for me when we got there. He put in temporary caps which made it possible for me to talk. The head-ache finally stopped. I felt completely drained from that head-ache. It had been a rough afternoon and I was pleased when we left the dentist's office. I had teeth, I could talk, and we were going home.

My teacher had called my house and spoken to my mom by the time we got home. She said that she was aware of what had happened and didn't penalize me for not being in class that day. She gave my mom her phone number and asked me to call her when I got home. I was tired and really wanted to lay down and sleep but felt that I needed to talk with my teacher first. I wanted this day to be over. I liked my Precalculus teacher but the day hadn't gone well and talking about Precalculus wasn't going to make it any better. However, the specifics of what happened in her class today would set the tone for my next class with her in the week to follow. I picked up the phone and called her. She said that the test in class that day had been postponed to the next week. The other students had too many questions and she didn't feel that anyone was ready for it.

She went over some vital information needed for that test and suggested that I study with Stuart or Stephanie. We concluded our call and I took her suggestion and called Stephanie. I didn't know Stuart very well and felt more comfortable studying with Stephanie. She and I met the next day at school and we discussed what was going to be on the test. She showed me some of the pointers our teacher had given us in class. All in all, it hadn't been a completely wasted day but my teeth have never been the same. However, the caps on my teeth have been there for almost thirty years now.

I've not encountered any another seizure since 1988. Seizures can affect my driving privileges. I was working at the time of that last seizure and they suspended my license for six months which is recommended by the Department of Motor Vehicles or DMV. A doctor has to sign a document stating that it is all right for me to drive every two years. This is a time in my life when I am most susceptible to losing a job. Either my license has been suspended and I can't drive myself to work or a seizure has happened at work, once, and it was too uncomfortable for me to remain working there after it happened. My dad couldn't drive me and himself to work so I tried to make other arrangements. Six months is a long time to go without driving and expect someone else to always be able to take me to work. This was crazy to me. Sometimes, whoever drove would drop me off at a city bus stop or a restaurant and I would wait there to catch the next bus going the rest of the way to work. This didn't bother me, as long as I knew beforehand that I was going to be taking the bus.

One time, when my license had been suspended and I had to drive twenty-five minutes in heavy traffic to get to work, I started driving two months before my suspension was concluded. There was no one to drive me and I had to drive myself or lose another job. I drove myself. I found out how important driving was when I started working. Otherwise, I could have used the city bus system and waited for a bus. Riding a bus to work will take at least one additional hour out of your day. You also have to take into consideration that a bus transfer may be necessary. It may take more than one bus to get you from point A to point B. Include one extra hour for waiting on the bus when working an eight hour day.

Driving a car to work reduces at least an hour of time wasted waiting for a bus. I like the city bus system because it gives you another way to get to work if your car breaks down. It isn't the perfect solution to transportation but it is efficient. Work is always quite demanding of your time and when your work day is complete, things can be done in a more timely manner.

I enjoy driving except for the monetary responsibilities of purchasing gasoline, vehicular maintenance, and ownership. Seizures are in my past currently and will hopefully remain there. Seizures kept me from pursuing any driving career opportunities until I ultimately succumbed to defeat. I wouldn't apply for a particular position if the classified advertisement suggested or implied that driving was included as one of the requirements. It has been essential for me to drive to get to work but that has been the extent of my driving career. Driving is a privilege and I have experienced this firsthand with intermittent driving suspensions due to the seizures.

My first seizure occurred after I started menstruating close to age eleven and one-half. I was diagnosed with diabetes at age nine and one-half and the seizures were aggravated by my hormonal changes and hypoglycemia. My pre-teen years were anything but delightful and the seizures seemed to abate nearer to age seventeen. Once I graduated from college, they became infrequent and required fewer and fewer visits to the doctor. Seizures were another element for me to contend with and have become non-existent over time. This is a chapter in my life that I wish to close forever.

Chapter 7: Diabetic Camp

During my middle school years, I was becoming old enough to attend a camp for medically handi - capped children like myself My mom and dad were looking forward to sending me there to meet others like myself with similar or the same medical problems. My neighborhood friends had grown distant and I needed to make new friends. Children were required to be at least eight - thirteen years old to attend this camp Going to this camp helped me and it was the second best experience of my life after being diagnosed. My tenth birthday was the first.

There were children and young adults there who ranged in ages from eight – eighteen. There were approximately seventy campers in attendance with diabetes, cystic fibrosis, cancer, chronic asthma, and assorted other disabilities. There were twenty-five counselors in attendance for the campers. One of the counselors, Darlene, passed away at age eighteen with cystic fibrosis and asthma my last summer there. Her lungs had filled completely with mucous from the cystic fibrosis and she drowned. She was having therapy a minimum of twice a day while I was there because I had seen her when I was giving therapy to an eight year old girl with cystic fibrosis.

Darlene was an extremely thin young lady at eighteen and her coloring was pale. She had left camp during the third week of the four week session. Everyone found out about what had happened to her our last day there. I didn't know her very well but she was very happy the last time I saw her. From what I remember, she had been planning on getting married at the end of that session. She was the only person that I knew of who passed away during a camp session while I attended camp for three consecutive years.

Three-fourths of the children and teen-agers attending this camp had diabetes. This camp revolved around structured activities for the entire four weeks. They had Arts & Crafts and there were so many things available there for children to participate in. Art & Crafts was usually held in the dining hall, as our group was quite large. The cabins were at the top of a hill where campers lodged before breakfast, lunch, dinner, snacks, and slumber. As diabetics, back then, we tested our urine four times a day. Blood glucose machines were not available for another three years. Urine tests were messy and inaccurate. You used a test tube and a dropper to test with. If your test was orange in color, your sugar level was too high. If it was green or navy blue in color, your sugar level was excellent. They preferred that mine run the green color because when it became that blue color, I was often too low.

Urine testing materials may not be available anymore in 2009. Mine were inaccurate and I was once told by a doctor that I may as well have not even done them. After having performed urine tests as directed for at least four years, it was disheartening to hear such information. I had heard about the blood glucose testing method and was looking forward to it's availability. I was tired of doing tests that weren't accurate and not helping me to control my diabetes.

We always went to the cabins before meals to test. At breakfast and dinner, we took our insulin. Some were taking insulin four times a day and made trips to the cabin sooner or more often than the rest of us. In between our necessary daily routines, we participated in the many different outdoor activities. They provided: horseback riding, archery, canoeing, hiking, tennis, volleyball, soccer, baseball, basketball, and a multitude of other activities. There was also a trampoline located next to our dining hall that we used all day long. There were lots of outdoor activities available to us. As campers and diabetics, most of us preferred participating in all the different activities as opposed to hanging around the cabins, until bedtime. When we went to the cabins, we had to attend to our daily routines by doing tests, shots, and working with our counselors to accomplish our personal medical goals.

I remember that there was one girl, Estelle, in my cabin who was afraid to leave the cabin. She didn't want to participate in anything or be around anyone except the camp director who wasn't located near us during our camp sessions. I felt really bad for her. I talked with her on her first day there. She was crying and I was a busy-body and wanted to know why. She seemed very sad for someone who was supposed to be making friends and having fun at camp. She had just been diagnosed with diabetes two months before entering this camp. She wanted to be with her family, instead of people who she didn't know. Her parents insisted that she needed to go to this camp to learn how to take care of her diabetes. I liked her a lot and attempted to become her friend.

Emotionally, she didn't seem ready to participate in what was being offered. I couldn't spend a lot of time with her while I was there because we were separated into different groups after breakfast to partake in their daily activities. However, for the first two days, she partook in the same activities as myself. We did our urine tests together and other cabin duties, since we were in the same cabin. Between having to take care of my own needs and administering physical therapy to an eight year old camper twice a day, it kept me industrious almost continually. Things became easier for her after a week but she was still unhappy. Her counselor called her parents at the end of that week. She was the only one that summer who left the session early. I felt that I had betrayed her somehow. My counselor, Vera, said that she was too sad to stay with us. I understood what she meant.

I enjoyed going to and being at this camp because I felt included in everything going on and by everyone there. I didn't feel like an outsider, as I sometimes did in a non-diabetic environment. I had to learn how to deal with my diabetes and this was definitely the place for it. I learned how to administer my own injections during my last week there. I was proud that I had finally been able to inject myself with minimal help from anyone else. Vera was training to become a nurse and would stand nearby me, just in case. It wasn't an easy task for me but I had finally succeeded in giving myself an injection. My parents were going to be very proud when they found out. All of the people who ran this camp were knowledgeable and truly seemed to enjoy being with all of us. It was a rewarding experience for me that has never been forgotten.

Chapter 8: Depression

Depression is a sickness, and diabetes is no different. Depression began to make an appearance in my life when I felt that I was unable to control my blood glucose, even after doing everything that was recommended. Controlling my blood glucose level was the only way that I could control my diabetes, according to everything the doctors had told me. I had been a diabetic for three years now, or twelve years old, when I recognized that depressing ideas were entering my thoughts. (ie: Why am I riding my bike for at least two miles a day if it isn't helping to decrease my blood sugars to where they should be as I was informed that it would? OR Why should I take that third injection of insulin today if it wasn't working like it should?)

I kept repeating the same recommended routines daily with minimal success. By now, I realized that I could never have complete control over the diabetes. I can always control my diet and exercise but not how and when my body is going to absorb the insulin that I dispense. Internally, my body had control over how and when ingested food was going to be metabolized.

I continued to practice the same suggested treatments year after year to control the outcome of my blood sugars, and the regiment for controlling diabetes hasn't changed dramatically in forty years. The medical technology has changed but not the treatment of the condition. The recognition that my meticulous treatments couldn't cure this disease and rid me of dealing with it for the rest of my life increased my depression. I continued to be optimistic and believed that there would be a solution to it's destruction one day. This solution to diabetes would set me free of administering insulin daily and end my fear of the repercussions from hypoglycemia and hyperglycemia. However, it appeared that my body was sabotaging my efforts to manage my condition and feelings of hopelessness amplified. This was when depressing thoughts entered into my mind and temporarily invaded my lifestyle.

I had seen a psychiatrist of sorts for three months, as my other doctor suggested, and he determined that I wasn't suicidal. I was too young to see a psychiatrist at that time and am not sure what his title was or his name. I was experiencing minimal depression because of the daily ministrations of my diabetes and it was to be expected. He had suggested that I involve myself in church and other social activities. This would keep my mind from wandering back to the diabetes and its unceasing issues.

I learned how to interpret the signs of depression to prevent the aftermath of those thoughts from emerging again. Thoughts of defeat and hopelessness make you feel as if everything you do is useless or worthless. I found that hobbies were a good way of daunting depression. I did things enjoyable to alter and improve my disheartening thoughts. Walking, bicycling, roller skating, and listening to music were just a few of my favorite pastimes. When I analyzed these activities, it was clear to see why I liked them so much. They were things that I began doing more frequently when I was instructed by my doctor that I needed exercise to control my blood glucose. Therefore, I enjoyed these activities before I became depressed and was diagnosed with diabetes. I had to dissect my ideas to better understand them.

Being depressed for a short while is normal but when it overtakes your life, it can alter methods established to manage your illness. As I got older, the time to do these things disappeared when money was vital and I had to work. Then, I discovered other things that were fun for me and included them into my life for relaxation and peace of mind.

Diabetes was an antagonist to me and I had to accept the fact that I was going to be sick interminably for the rest of my life. But there are ways to help me manage diabetes and medical technology is improving and making a difference in treatment. Getting discouraged when I am doing everything as recommended and am not seeing where it is making a difference doesn't help me. This is when I think back to everything I've eaten and done, for the past few hours or days.

What happened to change the result that I expected? If it was indeterminable, then it was something internal and possibly hormonal. I've been thinking this way for forty-one years. For me, this is important and makes more sense than chronically blaming myself for the fluctuations in my blood glucose levels. Infections can cause a spike in blood sugars also. It wasn't always my fault when this happened but, as I child, I believed that it was.

When I was growing up with diabetes, I learned a lot from my doctor. Sometimes, I would mention an incident to my doctor and he would say something that caused a light bulb to go on in my head. This was how I learned to deal with such an incident better, the next time. Most of the time, I was being taught about hypoglycemia and it's various symptoms and affects on me. I was athletic for a girl and hypoglycemia would surprise me sometimes. I had a vast array of symptoms from hypoglycemia. As these symptoms presented themselves, I learned to deal with them on my own.

My objective was to impede hypoglycemic shock and deal with the hypoglycemia while it was manageable. Hypoglycemia usually leaves me in a fog for a short while that can't be cleared until I eat something and certain foods work better, like juices, than candy. Learning the little things that can help to keep diabetes under control enables my day and life to run smoother.

Life was meant to be savored and enjoyed. Depression is disheartening and discouraging. This was another ingredient mixed into my ailment that I never wanted to share with anyone. Some of my friends at school asked me about it because they had heard something. I explained that it was the same thing as talking to the guidance counselor at school which was the truth. I didn't disclose the information we talked about, once a week for three months, to anyone except myself. I didn't write any of these sessions in a diary either, in case my brother took a notion to read it. Instead, I joined the book club at school and started reading.

I was able to recognize depression when it happened and immersed myself in all school activities imaginable. I had many talks with myself when discouraging thoughts entered my mind because I didn't want to talk about my diabetes or depression with anyone who didn't understand it. These thoughts weren't overwhelming. They were annoying and had an affect on other thoughts and ideas that I had. Me, myself, and I had some lengthy discussions about my depression created by my diabetes.

Depression is discouraging and …depressing. I don't want to be depressed and won't waste my time worrying about it. Worrying only makes it worse. Worrying also causes stress which can be detrimental to me. There were times, like a family death, when it was unavoidable but I wouldn't dwell on it. Life is too short and has too many wonderful things to offer. I don't want to waste any time being depressed over my diabetes either.

Getting involved in the different clubs at school established relationships with teachers and students. Some of those relationships led to friendships. It was apparent that my time was better spent reading or talking with my friends than focusing on how bad my life was. Everyone has something in their life that they prefer not to do. This is when procrastination begins and delays what needs to be done. Do it and get past it. Then there is more time for the things that you like to do.

Learning to deal with depression doesn't have to be dismal. The relationships established in book clubs and at church socials are important and can improve the quality of your life. If you are seeing a doctor, they may know of another group that appeals to you more than a book club does to me. Listen to them and find out more about community events. Joining a marathon group may be more interesting to you. Depression is a disease, for some people, and it doesn't have to be the motivating factor in your life. There are better things to do in life than concentrating on being depressed. The book club was the perfect fit for me, as you can see.

School Days

Chapter 9: High School

I had been looking forward to going to the high school that my big brother attended in our hometown. Then, I found out that my dad had gotten a job transfer and we were moving. However, when we moved, I determined that I could make some new friends there. This place was very country or rural, compared to the city where we had been living for the past fourteen years.

I found that I liked the people in this country town more so than those I'd left behind. They were more sincere and honest than most of my friends in my hometown. High school was a very exciting time in my life. We moved to this area during the summertime months. The boy who lived in the house before us, Nate, had been spreading 'rumors' about me to our neighbors. He told everyone he knew that a pretty girl was moving into his house. I thought it was nice that he would say this, since we didn't know each other very well. This small town had a recreation center where I met other teen-agers who would be going to high school with me in the Fall. I couldn't wait to get a schedule for high school, after being in middle school for the past two years.

There was a program that I participated in during my Junior year in high school that I really liked. This was a free program where I was tutoring an elementary boy in reading. He was in the fifth grade and reading at a third grade level. I usually helped him with his homework when he got tired of reading. I spent an hour with him twice a week. We'd spend at least thirty minutes reading and thirty minutes doing something else that he wanted to do. I enjoyed assisting him in his efforts to improve his reading skills. When we did the things he wanted to do, I found out more about what he liked. He needed the one-on-one attention in reading and it felt good to be involved in this program.

Recreational activities were my favorite pastimes like: roller skating and volleyball. In this small town, I had even gotten interested in bowling because my boyfriend, Roger, was on a bowling league. Seriously dating boys was a whole new experience for me. I was never sure how things were going to turn out with a boy, especially once he learned that I couldn't eat certain things. Some tolerated it and others avoided me like the plague. That wasn't any different than it had been with any other boy where we used to live. I didn't date or really like any boys when we lived in the city. Doing things on my own came naturally for me. All of my friends had a boyfriend so it was fun when I started dating Roger. He was nice when we first started dating. Later, he became more demanding and took me for granted, once we'd been together for six months. Things between us seemed to change with time.

High school was fun. I enjoyed dating, school clubs, church socials, and my high school prom. One thing was for sure: it was going to be another environment where I would need to be educating others about my diabetes. Once, I went to an entourage at a military academy for boys.I had never experienced anything like it before and it was different. This young man, Troy, didn't have a date for this entourage and his parents arranged for me to go with him. This was what broke up myself and Roger. I tried to explain to him that I was doing Troy a favor so that he wouldn't be the only one without a date to this event. Roger never believed me. I had to wear a floor length gown, not a dress, to this event which made it seem somewhat regal. Other dances that I'd attended with Roger were more casual but I still needed to dress according to specifications for that event. I was thrilled about being all dressed up and going to a military style ball.

Troy made sure that at 9:00PM that evening I had a snack so that nothing unfortunate could happen to me. He was very sweet but his parents had sent him to this school because he had been such a problem in public school. His school was located about two hours from where I was currently living. Our parents made sure that everything went according to how they planned it. He was dressed in the military academy uniform and looked stunning in it. He had on a navy-blue jacket with a red sash around his waist and white slacks. Black shoes and white gloves completed his ensemble. He was gentlemanly and handsome and I made sure that I told him so. I knew how important it was for him to hear these things in a situation such as this. My understanding was that his parents had sent him here to learn discipline and responsibility. He behaved in such a manner and I told his parents this also. I don't know what has happened to him since then. I have never been good at keeping in touch with people.

High School was an important time in my life, educationally and socially. My grades were not the highest in our graduation class but they were good enough for me to enter college the following year. My older brother wasn't as strong in his academics as I was. My parents wanted to send me to college and I wanted to go. My dad accepted another transfer back to our hometown and I went to the new community college to complete my first two years of college. I took the pre-requisites classes necessary there before entering into my major at our hometown university. My future was getting brighter with every passing day.

Chapter 10: College

College was a time for me to start making decisions about my future. My parents tried to talk to me about this but my mind seemed to still be on a high school level, especially socially, when I first entered college. I determined quickly that my college classes were going to be more difficult than my high school classes. They were more detailed and assignments were distributed on the first day of class. I attempted to take classes that I was interested in and that would assist me in attaining a higher than expected Grade Point Average. I discovered that there were classes being taught at night. A college environment was going to be completely different than high school. The teachers provided an outline which reflected the flow of the class for the entire six month semester. The outline provided us with a timeline for completing assignments.

Friends were an important part of my life during college particularly. If I missed a class, I would need to meet with a classmate for specific instructions for the next class. My school relationships before college were fairly typical. Even though my diabetes intimidated some of my classmates, some of them were my friends despite my disorder. I had lost many friends in elementary school and a new set of kids were around when I went into middle school. Because I had gone to elementary school with some of those same kids who were in middle school with me, word got around fast that I wasn't a popular child and was strange because of my diabetes or whatever reason they chose for me to be considered strange. I attempted to change this opinion of me with minimal success.

Some of the students hadn't liked me in elementary school but I felt that my chances were better for making new friends in a different and larger school. Middle school wasn't overpowering for me, as long as I stuck to doing my class work and homework. I became interested in Art because I thought that I liked a boy in that class and found that I wasn't very good at Art. My average grade in that class was a C. I aspired to make nothing less than a B in all of my classes. That meant that Art was added to my Busted or Do Not Take class list. Therefore, I decided never to take another class like that again just because I liked a boy in the class.

I continued to pursue my favorite classes in middle school: reading and volleyball. I took pleasure in English and Science courses too. I excelled in English and averaged above ninety percent of the students in my grade level. Volleyball had always been my favorite sport off of school premises. Dressing out to play volleyball in gym at school was a bummer. I perpetually played volleyball even when entering high school. It wasn't a complex sport to play and the team work required made it more fun. We moved to a different area during my high school years and other sports and subjects were introduced to me. Music or, more specifically, singing was my most alluring subject ever attempted. I was at every class no less than ten minutes before it commenced. As I progressed in school, there was a larger variety of courses to consider. I was well liked by almost everyone and this made me feel more comfortable in my school environment. I went out for volleyball again and was chosen to be on the school team. I tried to run in their short-distance track events but wasn't fast enough. Participating in hospital marathon walks to raise money for different organizations was more interesting to me.

When I got out of high school, my dad had taken a job back in our hometown. We moved back to that area after my brother and I graduated from high school. What I liked best about college was that I could sign up for the classes that I wanted to take. As long as they were one of my required courses under my curriculum, I would obtain credit for them. My dad was always trying to get me to take Math-oriented classes. He said that the English classes didn't hold a lot of potential for advancement opportunities in the workplace. I tried it but could only get to Pre-calculus because Calculus was too advanced for me.

Therefore, I kept taking English classes and began taking Spanish in community college. I had taken French in High School and found it entertaining and motivating. My French teacher was female and spoke with a German accent. My Spanish teacher was a male from Spain and spoke fluent Spanish including the accent. Spanish was painless compared to French. Then, I got into University Spanish where the class spoke only Spanish. My opinion of Spanish changed rapidly and I dropped the class after eight weeks. I couldn't pass any of the tests because our professor refused to speak any English. I understood several words in Spanish but wasn't fluent enough to pass this class. I really tried because I liked Spanish. When I dropped out of that class, I had to change my major as well.

I entered the University majoring in Paralegal Language Translation. I decided to change my major to Social Work. Social Work appeared to be similar to a Psychology curriculum. I was an English major and felt that this was what I wanted to pursue. I never went into Social Work but I took several higher level English classes that have served me well over the years. I have taken a multitude of collegiate classes and I believe that English remains to be my most revered subject to date.

I went to two different colleges; a community college and a university. College had been an extension of another year in high school and more classes to me. I was tired of school by the time graduation day from my High School arrived. My feelings are that I would have done better in college with a one year break from school after graduating high school. This issue is debatable, but not very for me. I worked as a substitute teacher for middle and high school children for two years after I graduated college. I was working in a clothing consignment shop near the end of graduation from the community college at that time and continued there for another year after my graduation. One of the teachers at the school where I graduated from owned this store and it is still in business today in 2009. I left the consignment shop in 1984, a year after I graduated from the university college. I wanted to substitute teach until I found a permanent teaching job with the county school system.

When I was substituting at a high school one day, a teacher who worked there suggested that I try to get a position with the County Recreation & Parks system. She said that it might be a stepping stone to getting into the School system. I had been unable to get a teaching job with the county school system but was able to get on with Parks & Recreation. The hours were part time and the work wasn't stressful. It was a closer match toward using my skills in Social Work than continuing to work in the consignment shop.

After six years of teaching after-school recreation which I truly enjoyed, I got married. I enjoyed the after-school program because of the schedule and the program orientation. The biggest problem with working hourly jobs is that they always interfere with lunch or dinner. This can create spontaneous situations that require special attention from me because my lifestyle revolves around when and what I eat. I never encountered a problem with scheduling when working in the after-school program. The hours were: 2:00 – 6:00P, Monday – Friday. I usually arrived at the center by 1:30 to construct stations for our planned activities.

However, salaries were minimal and it was a permanent part-time position. The only full-time position available in the childcare center was the Center Supervisor.

Center supervisors needed to be at their centers by 10:00AM to organize afternoon activities for the children who were enrolled in the program. For example, if we were going to a theme park, the center supervisor was responsible for calling the county school system for a bus. It takes a great deal of planning to provide quality day care to fifty to one – hundred children, every Monday through Friday.

I had wanted to supervise my own center for some time. There was a full-time center supervisor position becoming available in my center in the year to come. I wanted that position but knew that I was getting married and leaving the area. I was moving seventy miles away from that center to be with my husband. So, this opportunity evaded me and it's possible opportunities. I had strongly considered this position because it would be a full time position where I would continue what I was already doing part-time, with full-time status, better pay, and more opportunities for advancement. It was an opportunity that I had been looking forward to for the past five years and now it was here. I would have taken that position and my life would be much different than what it is currently if I had not gotten married.

We were relocating to the beach area and I had called to inquire about possible opportunities in their county recreational system for the summer. There was nothing available in that area at that time. My hometown county system operated differently than theirs and they started accepting applications earlier. I called past the time when they were accepting applications for the summer program and was unable to obtain a position. Marriage had changed my focus on life to include moving and living with my husband. Marriage changed everything for me. My life hasn't been the same since.

Chapter 11: School Experienced

I love school and going to classes. I just wish that, after all of the school and classes that I have taken, that I had truly been guaranteed a job afterward. Or, possibly directly before graduating college. I keep on struggling at fifty to find my niche in the workplace. Customer service jobs are where I have spent most of my time working subsequent to graduating college. I have been trained in Social Work, Computer Programming, and Pharmacy Technician fields and upon graduation, there is a recession and businesses aren't hiring employees. I will often take a customer service position that can lead to the career position that I want.

Until things improve economically, customer service is where I will remain. In my current working environment, advancement opportunities are less than minimal. Most of my co-workers wish to move up as well and competition is stiff. I want to work in a department in their store that doesn't exist yet.

Due to our economy, I have decided that I cannot take another class until my employer can compensate me for it. I have taken all of the classes that I can take in the field of Pharmacy Technician and wish to be working in a pharmacy. I was informed that I cannot get my degree as a Pharmacy Technician until I pass the National Certification Exam. I attempted to pass it last summer unsuccessfully and can't afford to take it again until I know that

I will pass it. I took most of my classes in the Fall of 2007 and wish that I had taken the exam in March 2008 the first time I scheduled it. I re-scheduled it in June because a classmate wanted to study for it with me. When I spoke to her about studying together for it, she had changed her mind. I was livid.

I am not certified and there aren't many uncertified positions available for a Pharmacy Tech. I will need to go in as an Aide and start at the very bottom. I feel that I can remember more if I am working in that environment and can hear what I've learned while I am there. Therefore, I am waiting to see what will happen next year. In my current position as a cashier in a grocery store, I have no one who I can ask any pharmacy – related questions.

College is diverse when compared to the last time I attended college in 2000. One day in the Fall of 2008, I got to school early to study for a test in my pharmacy class. I was sitting upstairs in an alcove, thinking that it would be quieter since the library was no longer located in that building. The college had built a new library located two buildings over from where I was sitting. I was seated in a bay window area studying and eating my crackers. I looked up from eating my crackers and was a little startled by what I saw. Two girls were standing next to the railing in front of me kissing each other. They seemed oblivious to their surroundings. Afterward, they separated and one girl left quickly and headed toward the stairs. This wasn't a peck on the lips either.

I was raised to believe that kissing is personal and can become intimate suddenly. Even if it had been a boy and a girl kissing, this kind of kiss was intimate and should have been done in PRIVATE. At least, this was how I felt about what I saw. I was a little uncomfortable but the moment passed. Just because I'm fifty doesn't mean that I can change THAT much with the times. I was taught to show courtesy and respect to other people around me, whether I knew them or not. I guess I'm old school and still believe in and practice moral values such as these.

I have a few more pharmacy classes that I want to take. They aren't required classes but they look interesting. I will wait until after I am employed in the field and can make more money to pay for my classes. With the current state of our economy, it could be awhile before I take another class. I've determined that I can obtain my pharmacy degree if I am working in a pharmacy with a pharmacist who can answer my questions. I need to pass the exam the next time I take it for professional and monetary reasons. The exam is complicated and will take a week or more to cram for. I have my game plan together for my next exam. The first thing I need is a job in a pharmacy.

I actually do have some advice about going to school. If a student has remained in school consistently for twelve years before graduating, they should consider taking a six - month leave of absence from school after graduating high school. College is an enormous commitment and you need to be ready for it. College is great if you know what you want to do when you graduate from high school. Waiting to enter would have been better for me. My parents wanted me to stay in school because of medical insurance reasons. I have diabetes and if I'd lost my insurance, the chances of another insurance company accepting me were questionable. If you are making this decision totally on your own, without any strings attached, take that necessary break from school after you graduate from high school. If you really want to go thereafter, then do so. After graduating high school, you have just finished twelve years of school non-stop.

School is a commitment that you shouldn't take lightly and, in my case, it's a lifetime commitment. I do believe in continued education in your chosen field of study. I know someone who has never been to college a day in his life. He is an artist who can draw almost anything imaginable. He specializes in drawing horror-type pictures. It depends upon what type of material he is using to sketch on. However, it is remarkable that he is age fifty and can retire and has never been to college. Had he taken art classes in college, he may have been able to retire earlier. Remember to choose wisely and choose what you want to do and what works best for you. Take advantage of the rewards going to school can offer you in whatever field of study you choose.

Trials and Tribulations of Marriage

Chapter 12: Marriage

This part of my book is awkward for me to write about because my marriage was estranged. Both of us had never been married before and we felt that we truly loved each other. But this was the most important time in my life involving my experiences with diabetes. It showed me incidences that I would have to face alone in order for me to survive. I loved my husband very much before we got married. Things changed quickly once we were together forever. Marriage was difficult for me. I knew that I needed to make my husband aware of certain things without scaring him. He already knew some but it is different living with diabetes than hearing about it over the telephone. He and I had a pre-dominantly long-distance relationship before we got married. We started seeing each other on weekends while he was stationed here on the East Coast for almost one year and our relationship became serious.

Patrick was the man of my dreams. He had dark brown to black hair with ice blue eyes. Everyone else calls it the Frank Sinatra syndrome. He was tall and carried himself well. He had a medium build; lean with maybe a few extra pounds around the middle like almost everyone that I knew. He was a military man and the only thing I didn't like about his career was that it would take him away from me sometimes. His career in the military could offer many benefits to myself, considering that pharmaceuticals were sold at discounted prices and food and clothing weren't taxable items on the military base.

He was stationed at the beach and I lived in a nearby city seventy miles from him. We spent weekends together and he called me every Thursday night to talk about what we could do for the upcoming weekend. We grew closer than we had been the last time he was stationed at the beach. We had known each other for almost ten years, on and off. When they re-stationed him somewhere outside of Virginia, we lost contact with each other. Or each of us was dating someone else when he came into town.

His mother lived about ten blocks from my house and he came to our hometown to see her and his brother and sister. Eventually, the military transferred him out to the West coast. His transient lifestyle as a military man kept distance between us the year before we got married. We saw each other four times in that year. I went there to visit with him twice when my center was closed during a vacation and he came home for Christmas and Easter weekend. Thanksgiving was one of the holidays I spent with him at their military housing facility for spouses. It was actually a motel that the military used for visiting family members with discounted rates and laundromat accessibility.

I liked California but my fiancé was homesick. In that ten year period before we were seriously dating each other, I didn't always see him when he was home. We were compatible and inseparable when we were together. We had been friends for a very long time before we decided to become more serious. I felt that this would help our relationship as it grew. This time, I didn't want him to leave without me. I was crazy about him and blinded by love before we got married and wanted to be with him for the rest of my life.

To this day, I wish we had eloped because tongues started wagging and our families wouldn't stay out of my plans for our wedding. Patrick and I talked about and knew what we wanted for our wedding but when the families got involved, our plans were…revised. Especially the mother and mother-in-law. My husband wanted to get married here on the East coast because they would re-station him here permanently once he married me. I resided here and the military attempted to keep married couples together. Now, I wish we had stayed on the West coast for some alone time for us. Staying here was not good for our relationship. He was on the phone with his mother frequently. I didn't feel comfortable involving my family in our personal affairs. I wanted us to create a life of our own. That may not have been the best idea in hindsight but I really wanted us to develop a lasting relationship and was determined that our marriage would succeed.

As individual people, we had a lot of similarities and differences in our lifestyles. Our similarities were small but important things like we both enjoyed ice hockey, putt putt golf, and Garth Brooks. These were things that we could do together and they were important, to me. There were other things too but those are the first to come to mind. Our differences were complicated. My differences were personal and revolved around caring for my diabetes so that everything else would remain in balance. His differences were more professional and it became personal after we were married and living together. We didn't physically live together until after the day we were married; June 6, 1993. He was a small weapons trainer for the Navy and used intimidation tactics that he brought home from work and tried to use on me. I knew that such tactics could be dangerous and I didn't want any such practices in our new home. I didn't see how these behaviors impacted us until AFTER we were married and living together.

With a wedding in our future, it was going to take a lot of time and patience for it all to come together on our wedding day. He asked me to marry him on Christmas of 1992 and I had already started planning our wedding months before he proposed. I was pretty much like any girl in this situation and already had a wedding dress picked out and my favorite flowers, in my dream wedding. Things turned out differently than my dream wedding. The year that he was on the West Coast before we got married, I spent countless hours planning our wedding. Both sets of parents lived near me and were constantly giving me advice about what to do for our wedding. I was thirty-three years old at that time and knew what we wanted. I loved them all and went almost certifiably nuts trying to keep the peace. I was so busy trying to keep the peace that I agreed to include things that I really didn't want to make everybody else happy. Big Mistake. Without my husband being here, his mother became somewhat over - assertive.

Her and my mother had spoken on the phone about something that upset my mother. His mother had said that the bridesmaid dress that I had chosen for my bridesmaids to wear would make her daughter look fat. I never heard about what she said specifically until almost ten years after our wedding day. My mom told me that she said something else and I wasn't surprised. My mom said that she never wanted to speak to her again. I just didn't appreciate his mother talking to my mother that way and I told Paatrick about it when he called me two days later. Once he spoke with his mother on the phone, things calmed down a little. This short reprieve gave me enough time to concentrate on other plans that needed my attention.

Because we were getting married on the East Coast, I had to re-establish myself in a church that I was a member of, once upon a time. Patrick came home one weekend and we went to this church together. I asked him if he liked it and he did. He himself hadn't been to church in our hometown for a decade or more. He attended the Naval church on the base where he was stationed at that time. So, we agreed to be married in the church in our hometown. There was one stipulation made by our preacher. When we talked to her about getting married there, she stipulated that I would need to attend that church every Sunday for the next year and my husband needed to attend church with me at least five times before we got married. Our preacher wasn't asking much of him but my commitment was much more intricate and demanding. I would be there for church socials and communions. I approached each Sunday, one day at a time.

Planning our wedding included: finding a wedding dress, choosing a caterer, choosing bridesmaids and the dresses, making bouquets and boutonnieres out of silk flowers for the bridesmaids and groomsmen, making bows for the pews where immediate family would sit during the ceremony, getting gifts for the bridesmaids, renting the reception hall at the church, choosing invitations and mailing them out. Oh, and I had to find a photographer. My list seemed to be never ending. I wasn't sure what kind of music to have at the wedding and used my boom box with a cassette tape for that part of the reception. That wasn't a very good choice but I was running out of time for planning that part of my reception by then. Patrick and I had chosen a favorite song between us; When a Man Loves a Woman by Michael Bolten. I bought the CD after we decided on a song for our wedding. There were several great songs on his CD and I used my boom box to play them.

My boom box couldn't be heard very well in the reception area but it provided background music during the reception. I remember that my photographer made a few announcements during our reception for the dance with the father of the bride and my first dance with Patrick as husband and wife. Our photographer also made the announcement when I was ready to throw my bouquet to the single women in the group of wedding guests. Since the reception was held at the church, there was no alcohol allowed inside the church and some of the guests weren't pleased with this. I guess I could have hired a Disc Jockey and had the reception elsewhere for those who wanted alcohol. It wasn't a total disaster but I could have done a better job of planning the reception. Next time, I'm eloping and no one will know I'm married until I get back from the Bahamas.

I had three bridesmaids and my sister was my maid of honor. My husband had three groomsmen and his brother was his best man. Our wedding party wasn't large but the little details were pertinent. I remember that on the day of our wedding, my dress was belled on the bottom half with the petticoat underneath. I was soo nervous that it rattled when I was shaking standing in front of the preacher. I wanted everything to be perfect that day. The most interesting stories about my wedding involved the bridesmaids dresses and the rehearsal dinner the night before the wedding.

During the rehearsal dinner, the night before our wedding, we met at an expensive Steakhouse and his mother was monetarily responsible for this event. My parents had offered to help pay for it a few months before that night and she declined their offer. The wedding party included: my mom and my dad, his mom and his stepfather – to - be, my older brother and his wife, my sister, my younger brother, my husband's sister and her boyfriend, his brother, myself, and Patrick, and his mother invited three of her friends and their dates to this event. The dinner was wonderful. We all had a salad with a baked potato, or whatever side dish we wanted like corn on the cob, and a steak. There was alcohol available and my parents and myself were the only ones not drinking.

When the dinner was concluded, his mother walked around the table and started collecting money from her friends, which was when my dad wasn't exactly sure what he was supposed to do. So, he pulled out his wallet and offered to pay for our meals. She declined, again, and my dad and my mom were very upset. Especially my mom. It caused some embarrassment for my family because we assumed that this was an event only to be shared among family members of the bride and groom. Friends were not included but I think she felt that she was paying for it so she would invite whomever she wanted to. It wasn't very kosher the way she handled collecting funds at the rehearsal dinner and I was about to find out more than I wanted to know about her. His brother and sister were more subtle than she was. I didn't know much about his new father-to-be and wasn't too concerned about him at that time.

The situation with the bridesmaids dresses was unbelievable, to me. I was told by my mother that the dresses that I had chosen for my bridesmaids to wear were inappropriate. She was adamant about this and I wouldn't talk to her for almost a week after she made this announcement to me. She asked my sister to talk to me, instead. My sister said that mom wanted to do something for my wedding, since I was doing everything, and offered to pay for the dresses. I was extremely upset and didn't want her to pay for the dresses. I wanted them to wear what I had chosen because it matched my wedding gown. I had chosen a suit with a matching jacket and skirt of different colors; pink, peach, and pastel blue. My sister's maid of honor dress was mint green. My mom wasn't going to take NO for my answer to her suggestion. So, my sister was over our house every night for a week trying to convince me to let mom make the dresses for me. I unwillingly conceded and regretted it from the moment I let her intervene. I was very unhappy about this decision but was tired of fighting about it.

My mother took over this duty and the dresses were two sizes too big for all of my bridesmaids. I found out years later why she was so bent on me changing my selection. His sister had called his mother and cried because she said that the suit I chose for them to wear made her look fat. So, his mother called my mother and told her how her daughter felt. Mom was not happy about having to lie to me and told her never to call her again. They saw each other at the wedding but I was distracted during my wedding ceremony and didn't notice that they literally weren't talking to each other. They never spoke again after that happened.

Before we left the church, I went into the changing room and changed out of my beautiful wedding gown and into a gorgeous white suit that I had found in a nearby department store. The jacket and skirt were both white and the jacket was double-breasted with gold buttons down the front of the jacket. Patrick was still wearing his black tuxedo. He had driven his Ford Ranger truck to the church and we left in that truck. But as we were leaving the church, everyone was throwing rice which is traditional at weddings. It collected in my hair and trickled down the front of my jacket and Patrick's hair and tuxedo were full of rice when he changed at home. We had driven to his mom's house after the wedding to change into more comfortable clothing. There were a lot of things that happened that day and it is hard to express all of them. However, it was the most beautiful and wonderful day of my life.

While we were at his mom's house, we collected some of his things and put them in the back of his Ford Ranger truck. He was used to a transient lifestyle and knew exactly what he wanted to take with us when we were moving into our life together which was soon to begin with our honeymoon. We headed for my house after that so I could change and make sure I had everything before we left to go on our honeymoon. It was a blessing that our parents lived close to each other because this moving process took us about thirty minutes. If you had asked me this question BEFORE we got married, my answer would have been the opposite of this. We spent the weekend at our most beloved hotel because of the luxurious accommodations and its location. We knew what to expect from the hotel and staff which was preferred over the previous year of my life. Patrick liked the idea because he had driven from California to Virginia and needed a reprieve from all the driving. His truck was not automatic and I could drive it for long distances but that was all that I could manage.

We had decided that we would take our real honeymoon together when we moved him back to Virginia permanently. He was stationed in San Diego, California and we were going to drive from there to Portsmouth, Virginia. The most memorable part about our honeymoon was when we stopped at the Grand Canyons. The massive expanse of rock and cliffs was amazing. And his picture with the gorilla was hilarious. It was going to be a long drive but we were going to be together and I couldn't envision anything better than finally being alone with him for one entire week. Our honeymoon had lasted more than one week but we spent most of it moving.

When we relocated and moved all of our belongings to Portsmouth, we lived in an apartment for our first year together. It was located near a mall and we had ample places to shop if we needed anything. The apartment was a three bedroom structure with a galley kitchen, our own washer and dryer unit in the hallway, a kitchen nook, one full bathroom, a den and living room area. There was also a sliding glass door that went from the den to the patio outside. We were located in the downstairs apartment and Patrick got tired of the noise coming from our upstairs neighbors within a month after we had moved in. I liked walking to keep my weight under control so I walked around the complex twice a week. I enjoyed living there, until it became obvious that we had more things than what we could fit into that apartment. So, we started looking around our area for something with more space.

We found a lovely one-floor town home that we liked the following year and didn't renew our lease with the apartment complex manager. The townhouse had two bedrooms with a full bathroom in each bedroom. The bedrooms also had walk-in closets, which I loved. There was a door leading out into the back yard from both bedrooms. And there was a galley kitchen, a kitchen nook for our table and chairs and a den and living room area. Our washer and dryer was located next to our kitchen behind the French doors in our hallway. The bedrooms were much larger than the ones in the apartment. We also had several more closets and the walk-in closets for storage space. Also, there was a shed in the backyard where the lawn mower resided. We had yard tools stored in that shed and I believe there was more artillery being stored there too. I will go into more detail about this later on in this chapter.

There was a beautiful Victorian house that we'd seen and that I liked better but I didn't want to persuade Patrick to buy something that would over-extend our finances. I really wanted the house but changed my mind after discussing it further with Patrick. So, I settled on this lovely town home. We signed our mortgage papers to the townhouse and moved in one week later. The apartment had been nice and the townhouse was going to be even better.

During our marriage in that beachfront area, I found that I didn't know anyone there except for the people I came into contact with at work. My husband, however, knew some long time friends of his from when he was stationed in that area before. We hung out with two friends of his who were married and lived in that area and who I'd seen at our wedding. They were nice but they were friends of his. My friends were nice too and there was one couple that I wanted us to get together with. However, her husband was re-stationed and they moved from our area to the West Coast. We were unable to meet somewhere before they left. I was disappointed because I wanted him to meet with some of my friends so that we could socialize.

Most of my other friends were either separated or going through a divorce. They knew I was married and I tried to communicate with other married people that I worked with. I started to feel like the outsider because we were always going out with his friends. However, being the outsider was something that I was familiar with. I was used to this role after living with diabetes for twenty-five years. And then, we began to have some problems. Anytime we went out, it was always to meet with his friends.

I had been dealing with the diabetes primarily on my own for twenty-five years with my family's help, especially financially. It wasn't that I didn't think people had ever heard of diabetes before. But hearing about it over the telephone and living with it are poles apart from each other. Seeing the regiments necessary for maintaining control of my condition and hearing about it over the phone are diverse because once you hang up that phone, you don't have to think about it again until you are with me. The disease effects each individual in a specific way. It is treated differently, depending on if you are a type one or a type two diabetic.

Type one's, like myself, are insulin dependent continuously. Type two's can control their diabetes using pills and exercise to keep themselves from needing to take insulin by injection every day. I remember explaining this to him, hoping that it would make some kind of sense to him. I was already ahead of him by twenty-five years with my education and knew that it would take him awhile to adjust. It took my entire family and myself two years to fully understand and adjust to it. Unfortunately for us, we didn't make it that far.

I don't think he ever understood it and maybe even tried to deny that there could be possible problems with it. This was my way of reasoning through it when I was first diagnosed. Denial tactics were a part of my life from age nine – ten. If I didn't have it, I wouldn't need to take the shots. That was my thinking process at age nine-ish. I didn't learn how to give myself an injection until I was eleven years old. My mom had been unable to do it also and never attempted it again after I began administering them. Learning to give a shot is a part of my memory that has been blocked for some time because it was so unpleasant.

I remember that after I learned how, my dad didn't give me any injections ever again. He had given my injections to me for two years and it's been my sole responsibility since then. My dad and I lost that bond with each other when he stopped giving me injections. He was giving me the shots because he knew that I needed them and mom couldn't do it. It was essential that he did it at that time or he would have needed to hire a nurse just for that purpose. Dad passed away in 1996 and if I need an injection and can't do it myself, it will always be a 9-1-1 situation. I can't see why this would ever be an issue for me but one never knows these things.

My ex-husband never learned how to give injections. My doctor asked me if he knew how and I told him that I wasn't sure. I never brought it up to Patrick and he never asked. We weren't together long enough for him to take a part in this anyway. I had been dealing with it myself for twenty-five years before we met and it was always my responsibility, just like it was most of the time with my family. There may have been an incident where I may have needed him to do this but nothing like that has ever happened to me before. I wouldn't have seen him as weak or blamed him for not being able to do it when my own family members were unable to. It took me two years to learn how to give myself an injection. I don't expect other people to know how to do this. I've had diabetes for forty years now and giving an injection is like riding a bike for me; you don't forget how to do it after you've learned.

I was on the injection while we were married and they were given at breakfast and dinner, religiously. I made sure that I was doing a lot of walking and walked around our apartment complex and then our town home complex at least twice a week. I went out dancing, usually on Fridays. Where I was living then, Friday night was always the best night to go out dancing. Sometimes I missed Friday night dancing when I was working. So between the two days walking and the one night dancing, I was keeping my weight under control and managing my diabetes. Weight management is an important part of diabetes. If you can keep your weight within a normal range, your body is processing food correctly. If not, your treatment may need to be adjusted.

Diabetes is a constant balancing act between food, insulin, and activity. Everyone has some sort of dietary issue that they must address which makes food significant to non-diabetics as well. My mother needs to watch her sodium intake for better heart health. I keep trying to manage my diabetes but getting older deters some regiments from being effective. Unfortunately, I can't stop myself from getting older any more than I can stop myself from being a diabetic. If there is ever a cure for diabetes, then there will finally be a change for the better in my life.

Marriage was a hard road for me. I remember when my husband and I spoke about diabetes that he mentioned his grandmother on his mom's side of the family had it. We went to visit her on Thanksgiving that first year that we were married.

She baked a turkey, had green bean casserole, sweet potatoes, corn, rolls, stuffing, and a few things that I couldn't name. The rest looked like dessert to me. So, I had the turkey, corn, rolls, a little stuffing and some of the green bean casserole.

There were some foods with sauces that I also avoided. Sauces have hidden sugars that I've been warned about since age nine. His mother stood up for me and said that I always ate like a bird. (I was the thinnest person in the group). I told his grandmother that everything was wonderful and that I was just trying to be careful. She understood that, as she was a diabetic herself. I don't eat many dessert type foods anyway unless I am hypoglycemic. Then, I have to be careful because I have a tendency to eat too much sugar. This is when my diabetic knowledge and practices are most evident.

Patrick received a rifle during our visit there and this took my attention completely away from the food. His grandfather had passed away earlier that year and his grandmother had saved it for him. I was unhappy about this. We lived in a one-floor town home and already had two guns in the house that I felt were unnecessary. Now we were going to have a third gun in our midst. I realized that this was a big part of what my husband was doing in the Navy and that's why it didn't bother him. He was a Small Weapons Trainer. The fact that I didn't like them seemed to be a mute point to him.

We took the rifle home and argued about it all the way from Light Foot where she lived to Portsmouth. I got tired of arguing about it and didn't talk to him for that last half- hour before we got home. As soon as he walked in the front door of our town home, he put that rifle in the closet directly behind the front door. We already had a pistol in our bedroom in a nightstand next to our bed. He had another gun in his truck. We were running out of room to store his artillery in our one-floor town home.

Human nature seems to bring out the worst in us, sometimes, and we have a tendency to remember all the bad things before we remember the good. Our marriage was wonderful for the first year we were together. We went to dinner by ourselves and with his friends, Ethan and Winifred, most frequently. We went to play putt-putt often because it was directly behind the apartments where we had lived. I wasn't into it as much as he was and wasn't as good as he was. I didn't dislike putt-putt golf. I grew tired of it. We started going so much that I knew there was something else that we could be doing together.

We went to a military Christmas party, once, where everyone had to dress formally. I wore a nice dress and he wore his navy blue uniform. He looked sharp. They weren't required to wear the white gloves for this event but the uniform definitely made a statement. All military personnel wore their uniforms and everyone else, like myself, dressed in evening gowns, fancy dresses, flashy suits, and tuxedos.

There were many things that we had enjoyed doing together that seemed obsolete after our first year together. Things seemed to be moving so fast. I felt that it would take some time for us to enjoy being with each other the way we were before we got married. I didn't know how but I knew it must be possible. We hadn't found a church in the area to attend yet and this was something we had both wanted to pursue. I began asking my friends at work where they attended church and most of them weren't attending a church. It was going to take me some time to find a church where we both would feel comfortable in attendance. I really wanted to do this and I enjoyed the church services and the people. There were many churches in our area and I was determined to find the right one for us. Unfortunately, time was not a luxury that worked in my favor.

Chapter 13: Children

Children were never meant to be a part of my life. However, now that I was married, I wanted to investigate this aspect further. I had been told all my life since age nine that I should never have children. I didn't completely understand why but my doctors discouraged me from having children. They said that it would be too tough or taxing on my vital organs for me to carry a baby. My body was pretty thin up until about age forty-five. I think they thought my body cavity might be too small to carry a child successfully full-term.

Therefore, I had considered adoption, even before we got married. However, for health reasons, it didn't seem like a good idea for me to become pregnant. We only discussed children one time in that year and a-half that we were physically together and we never broached that subject again. After what had happened between us, I wasn't sure about having children with him anymore. I was definitely open to adoption but never had enough time with him to know what WE wanted. One thing was certain: I was open to adopting a child with him if we could work things out between us.

I sincerely considered adoption while I was with him. I didn't talk to him about it but I thought about it, sometimes. I was pretty sure that he wanted a baby and I was having problems understanding or getting past some of our difficulties. I couldn't always see his emotions in his eyes. This may have occurred from his training in the Navy as a Small Weapons Trainer. Their tactics were to intimidate your enemy and overtake them. The issue with this thinking or reasoning is that I wasn't his enemy.

I learned that I wasn't very good at showing my emotions to him either and felt that this may have been due to my diabetes and my practice of hiding my feelings about my symptoms of hypoglycemia. Stress reduction was included in my therapy every day and I used methods of hiding my emotions to avoid stressful situations. It was obvious that we both had different methods of hiding and showing our emotions. I didn't believe in divorce and I don't believe in violence. I would have done anything to keep us together. Finally, fear got the better of me.

I am a health-conscious person and, especially when I am around other people. I am not perfect, as I have mentioned before, but I try to pay attention to what I am eating and drinking. Alcohol is too dangerous for me so I am not an alcoholic beverage drinker. I tried it some in college and stopped after about six months. It increased my blood sugars rapidly and to the point of being almost critical a few times. I was able to get it under control before it became critical. There are so many things that need to be addressed during pregnancy and nutrition is on the top of the list. Because we never discussed this subject again, I wasn't exactly sure what he wanted and what to think. He would have needed to be clear with me about this if he wanted a child.

We never got far enough along in this discussion for me to approach my doctor. I never got into any specifics with him because I believed that he was against me conceiving. I knew that my doctors in our hometown were against it and believed that my new doctor would be too. I enjoyed children and had taught children up until the week before we got married. I've never had any children of my own. Now, since our relationship ended in divorce, we'll never know what a blessing life could have brought us.

Chapter 14: Divorce

I believe that my ex-husband and I would still be together today, fifteen years later, had he not hit me in the back with his fist which was an intentional act of violence. I loved him and couldn't believe that he would have ever done anything like that to me. That one action changed my opinion of him and altered the course of our future together. I believed that once you got married, you were married to that person for the rest of your life. I was in our relationship until death do us part. That time in our marriage has several missing pieces that can only be found by him to complete our puzzle.

The divorce was harder on my body than it was on my mind. Initially, I was able to handle my stress levels without encountering any disastrous hypoglycemic episodes during our separation. He was able to handle my medical finances and I was able to live my life without the fear of another altercation with him. Once my savings were gone and my funds were running low, then I was told that the department store would be going out of business by the end of the year. They say all good things must come to a close but I guess this can be true for bad things too. Losing my job caused me to have to look for more lucrative job to career opportunities.

I tried to keep my blood sugars under control but they were dropping too fast and I wasn't exactly, without a doubt, sure why.

I knew that stress could cause this to happen, as well as over-activity. Between moving and ALL the responsibilities associated with that, my responsibilities grew and my paycheck kept shrinking. It involved keeping an eye out for my new roommate and her friends, finding a new job and getting prepared for interview(s), keeping a close check on my medicine and diabetic supplies, and trying to adjust to a whole new lifestyle. (Oh! And don't forget going through a divorce and having to pay my lawyer every month). I assumed that it was stress and when I went to visit my doctor for an annual check-up, he prescribed …..PROZAC which is for STRESS.

I felt fine for the first two days that I took it. I continued to take it for one complete week until I started feeling not quite right. When that week ended, I stopped taking it. I found out later that it didn't mix well with one of my other medications and that's why I was feeling strange. I decided that it was something that I didn't need so it was no great loss. Also, it was expensive and I couldn't afford to deplete my funds with an unnecessary purchase. I needed insulin and the other medicines more. I should have seen him again to determine an alternate medication for stress but I decided against it. My funds could not support another expense.

I initiated the separation one weekend when he left town. I had been afraid for over six months and took that first step when I felt it was safe. By that time in our marriage, I didn't know what else to do. Things started moving too fast for me and my lawyer had a lot to do with this. I wasn't exactly sure that I should have initiated it but I felt that I needed to do something before things got out of hand again and he got home from his trip to his mom's. He said that he was going to visit his mother and I believed that he was but couldn't be sure. We had become estranged and hadn't been talking to each other very much. My fear had gotten the better of me and I felt as though I had been forced to do this. I didn't believe in divorce and was crushed when I realized that all other options had been exhausted and I was ending our marriage.

I told myself just to take it one day at a time because it was so overwhelming that this could happen to me. Once I had initiated our legal separation, I wanted the divorce finalized before I decided what my next move would be. I didn't know what the court would award or retract from me so I had to wait it out. We met at a commissioner's office to finalize our divorce exactly one year later. I was totally broke and couldn't even afford to pay the commissioner so he set it up and paid for it. He had set that appointment and I made sure I was there. I told the commissioner what had happened and he ruled it as irreconcilable differences.

My ex-husband had hit me in the back with his fist, and his hand is three times bigger than my own hand. I lost fifty percent of the functioning of my right kidney with that punch on that day. He never made an advance on me like that again while we were living together. I made sure of it. We had been to the Navy Social Worker and he called us back once about a week after we visited with him and left us a message. We were supposed to meet with him again, two weeks later. I tried to call him back three times and never got a response from him. It seems that our case had been dropped.

I knew we needed someone to talk to about what had happened and with no help from the Navy, I started looking more diligently into neighborhood churches for couples counseling. My husband never offered any suggestions for reconciliation. We weren't talking much at that point and I didn't want to send him into some kind of a rage and have something worse to happen. Going to the churches seemed to be the best answer to our situation.

He was treating the situation as if nothing ever happened and talking to some married woman on the internet about our marital problems almost every night. Back in 1994, I didn't know anything about the internet. All I knew was that when I tried to make a call to one of my friend's when he was online, it would disconnect him from the internet. He appeared to be trying to isolate me from my friend's but he felt that it was ok for him to be talking to that married woman. There is no telling who he was talking to on the internet. I told him that I didn't like him talking to her on the internet about us. He didn't seem to care.

This was a hot button issue for me and I let him know it. I felt it was important for him to know how I felt about it. We talked about her several times and he seemed to believe every word that she said. I was frustrated because she was driving a wedge between us and I couldn't see her to tell her this. He kept communicating with her on the internet and it seemed that there was nothing that I could do about it. I was disappointed in him and felt helpless to do anything about it. I was not computer literate, at that time, and did not know what I could do to stop him from communicating with her. I wanted to take a baseball bat to the computer but was afraid of the repercussions.

My goal was to find someone for us to talk to about what he had done to me. I had become very quiet around him and started keeping to myself. He took me to a church meeting once and dropped me off, as if I were supposed to be at this meeting alone. He didn't offer to join me at this meeting. It was a battered women's group and there were two husband's in attendance with their wives at this meeting. I spoke to three different preachers at three different churches and tried to encourage him to go with me. He didn't express any interest in this idea.

I was extremely frustrated because I didn't believe in divorce and had never wanted to see that day come. I knew I couldn't deal with him physically by myself, if anything happened and that is why I sought help outside the house. It was a very strained time for me and my blood sugars were just starting to become affected by it. But I couldn't improve this situation on my own and knew it. Anyway, once I left him, my stress levels seemed to improve some for a few months.

We were very happy for the first year of our marriage. We did small things together that all couples do. We went out to dinner and movies and even went to a Garth Brooks concert at the concert hall, once. We met his friends at the Outback Steakhouse one night for dinner. He found out how much I liked to shop. I did a lot of window shopping, mind you. I wanted many things for my new husband and knew that our living area would get cluttered quickly if I gave into this idea. You have to remember my mindset: I am not a violent person and would have done anything that I knew of to make him happy. When we were first married, I was more starry-eyed and gullible. I have come back from being unconscious from hypoglycemia before and have faith that I always will. Faith is something that I have felt and have deep inside me. Even though I've had a difficult time getting to church because of working on weekends, my faith has always been with me.

I don't know why I ever thought that a marriage could work out between me and anybody because of the demands placed on my life by my illness. These demands are written in stone and can only be manipulated when the situation warrants it. I have to be aware of my symptoms before I will take action. Eating something sweet when you think you feel hypoglycemic can make matters worse, if you are already running high. Any changes made must be done with the utmost care and consideration of previous foods ingested, insulin taken, and activity regimens. When I tried to explain it to him and the importance of certain restrictions, I probably wasn't able to adequately express how vitally important it was for me to adhere to them.

Diet, Insulin, and Exercise must be included in my lifestyle until death do us part. Diabetes becomes a lifestyle because it is my commitment to keeping this disease from killing me. He knew that diabetes was serious because of his grandmother but had never lived with it for any length of time before. It is a commitment that can't be broken. The consequences are dire. I will remember this, if I ever fall in love again. It may never happen for me, again.

I continued to struggle financially and started looking more relentlessly for gainful employment. I was in a good place, before I was married, to run my own daycare center through the county. However, my marriage took me away from that area and when I made the move with him, I checked into this before I resigned. I went into a new area and found a job in a department store. I was satisfied with the job in customer service but not the hours. My schedule was unpredictable, unlike my husband's. He had 2 late weekend nights a month for what was called a watch and the rest of his hours were from 8:00 – 5:00, Monday through Friday. When I left him, I roomed with my friend and her high school age daughter. Natalie and Karen, successively. Natalie and I struggled with finances frequently when we got together but we had our freedom.

Her husband of twenty years threatened to shoot her and that was all the encouragement she needed to leave her marriage with her daughter in tow. He was a drunk and a mean one, from what she said. She was much better at handling abusive situations than myself. She was also physically larger than myself. I was tiny most of the time and remained that way as much as possible. Going to the doctor included consulting with him about this was a part of my medical treatment at least twice a year. So, I started going dancing on Friday and Saturday nights. I wore my engagement ring when I went out because I didn't want to be bothered until after my divorce was finalized. I didn't want any complications. I continued with my exercise regiment and met with my friends from the department store when I could which was where I met Natalie.

The line between the time when I was married and then divorced will overlap at times. But my story doesn't change. I'm glad that our divorce was finalized. Dealing with whatever was going to happen next in my life was causing havoc with my blood sugars. Moving on with my life wasn't an easy matter. Getting a good job in that area seemed impossible since I wasn't a military employee. This particular area was regulated by the government and it's employees and I am categorized as a civilian. Things got harder and harder for me and my friends stuck by me one hundred percent.

Two years after I left my husband, I was still residing in the area and met someone who made me happier than Patrick had. I liked him but was still scared and it took me some time to trust him. After dating him for six months, I realized how sweet he was. I appreciated the creative side of his personality because I had to be creative continually. He was an architect. It was obvious that he enjoyed his work, just not necessarily the people he worked with. I had a true understanding of that too. Some people, I really like to work with and others, I will work with but not voluntarily. It is in my nature to be helpful whenever possible. I will always be the last one standing, in any situation.

The diabetes has brought me down a few times but I constantly seem to bounce back. I believe this is true, as it has happened several times and I will prevail. This is part of my faith in myself and my abilities to prevail over my diabetes. My husband moved someone into our town home with him one month after I left him. He obviously moved on immediately after I left him. He was probably with her before I left him. He had been pretty sneaky about it because I never noticed that he was seeing someone else while we were together. It was obvious the day I saw another car in our driveway on my way to work.

After the divorce was finalized and I had lived there for several years afterward, I recognized that I was never going to be able to make enough money to survive there. So, I moved back to our hometown and found a job doing temporary work. By temporary I mean that I worked for a temporary agency. I tried to get on permanent with that company and found that they didn't want to pay the finder's fee to the temporary agency who employed me. There was a $1000 mandatory finder's fee if I would have been hired permanently. I loved that job. It was similar to what I am doing sitting here in front of my computer, except that you have to deal with a continuous stream of phone calls. I worked in a call center and it was pretty peaceful, until I had an irate customer on the phone. Many times, I was able to handle those calls successfully. However, there was that occasional caller who had more severe issues and wanted to speak to a manager.

Working with a customer on the phone is easier than face-to-face interactions. Working with them on the phone is usually faster and more efficient particularly for the customer. Anything they need can be determined more quickly than driving to the bank and waiting in line and then, finding out that you needed to call another division of the bank nonetheless to get your question answered. Waiting on the phone can try your patience but so can sitting in your car at a drive thru window.

This temporary assignment ended after one year. I obtained other assignments through the temporary agency working for other companies frequently. I wanted something permanent with benefits because to me, temporary meant temporary because of that finder's fee. This meant that jobs would come and go and I wanted to work in the same place consistently. To this day, I am still eluded by that perfect job. Either a company is not hiring, the application has to be filled out online and I am not sure which store will call me, or the economy is so bad that upgrading our store has been postponed until it improves. I knew that I wanted to move on and all I was getting were excuses as to why there was a hiring postponement or freeze. It was always something.

My divorce destroyed the delicate stability of my life, alone. I was left trying to put the pieces of my life back together. My personal life was resolved instantly and involved the assistance of my immediate family. My professional life was not as simple to settle. I didn't feel comfortable returning to daycare with the county after only being gone for three years. With the multitude of people that I knew in that system, it felt overwhelming to me. Many of them were my friends and would be asking a lot questions that truthfully I didn't have the answers to and wasn't ready for. My professional life was shattered and keeping me from moving ahead. My funds were minimal and left nothing for extraordinary circumstances. I was living paycheck to paycheck. If there ever is a next time for me, there will be a nest egg stashed away specifically for emergencies.

Hypoglycemia Road Blocks
During Divorce Proceedings

Chapter 15: JailBait

I was living in the Virginia Beach area during my separation. This chapter is about what happened while working in that area on a hot summery day in May. I got up on that hot, sunny, summery day in May and got ready for work. I wore a navy blue floral skirt with an off-white, short-sleeve crew neck top and a navy suit jacket. WITH my black pumps and purse. I remember this day like it was yesterday. I brushed my teeth and put on my make-up. I was living in an apartment with Natalie and Karen, at that time. I was on my way out the door and remembered that I hadn't eaten breakfast yet. I grabbed an apple and headed out the door.

When I got to work, I had a cart piled high with papers waiting for me to deliver to different departments in the office. The documents were delivered to the table where I worked and then it was my job to organize and distribute them. This was a newspaper job that involved secretarial-type duties including loads of paperwork with sorting and distributing mail being #1 on my list of responsibilities. I got to work by 8:00AM and only had an hour to peruse through all that mail. I tried to ask for an assistant to help me labor through that mail and they always responded that someone else had already done that. That's how I ended up with ONLY one cart load of papers to go through that morning.

I kept struggling with it and, on that May day in 1996, I decided that I needed to go home for lunch to get out of the office for a change of scenery. My apartment was ten minutes away and I had an hour for lunch. I was having problems concentrating and focusing on what I was doing that morning anyway and needed that break. My blood sugar was dropping and I didn't realize it because my mind was preoccupied with mounds of undistributed mail.

This made it impossible for me to complete my job before lunch. I decided that I needed the break and felt that they could wait until I returned from lunch. When I left the parking lot for lunch to go to my apartment, it was approximately three turns from where I was parked. I was driving a 1991 white, geoprizm car then and it got me everywhere with minimal maintenance and gasoline. When I pulled up to Virginia Beach Blvd., I made a right turn into the far right - hand lane and kept straight. The road into my apartment complex was ten – twelve blocks from there. As I was driving down Virginia Beach Blvd so that I could make a right turn into my complex, a city bus got in front of me. I didn't wish to remain behind that bus because my visibility was impaired by the bus. An opening became available in my left-hand lane and I glided into it without incidence.

Traffic was congested and it was hard to see around the bus. This highway was a four-laner and two cars got in front of me when I pulled into that lane. I still felt that it was possible for me to pass that bus and get home for lunch. By the time I got to the traffic light on the block where I needed to make that right turn, the bus was directly beside me. I was having trouble thinking and was too late to pass it. Traffic was heavy and I had to keep driving. I was stressed because I missed my turn and had not completed my work. I had one big pile to organize and then it would be ready to distribute.

On top of all that, it was very hot and the air-conditioning was broken in my car. Despite the heat and the fan in my car, I was feeling shaky and weak and suddenly remembered that I hadn't eaten very much for breakfast that morning. That just made me more nervous. But all I could do was focus my energy and attention on the road and the cars in front of me. I tried to get back into the right- hand lane several times and kept getting cut off. Knowing that I wasn't feeling well made me more wary of changing lanes and hitting another vehicle. All I could think about was going home to get something to eat. So, I kept driving.

Finally, my blood sugar had gotten to a dangerously low level and my mind was going in and out of consciousness behind the wheel. I drove fifteen miles in that condition before I stopped at a stoplight with my foot on the brake and passed out from hypoglycemic shock. Someone must have called 9-1-1 because a policeman arrived on the scene first. I never saw him. I can't remember anything that happened while I was unconscious. An EMT, Emergency Medical Technician, described what happened to me before they arrived when I became coherent.

She didn't give me all of the details but I comprehended the important ones. She said that the policeman was going to take me to jail if they had not arrived when they did. He had my car in park, my seatbelt unbuckled, my driver's door open, and I was slumped over the wheel. He was going to lift me out of the car and put me in the backseat of his police car. It was obvious that he hadn't checked the stainless steel chain around my neck that identified me as a diabetic. I wore it every day when I lived in Virginia Beach. Had the ambulance not gotten there when it did and helped me, I would have been ..... JAILBAIT. And had I been put in a jail cell, I might have died from hypoglycemic shock. Fortunately, the ambulance was there approximately one minute after the policeman. She, Sofia, said that she had to ask me three times if I was a diabetic before she could get a response from me. I WAS GONE.

This is what happened AFTER they got there. When she got there and saw what was happening, she asked the policeman to move so that she could ask me some questions. He unwillingly complied. He was convinced that I was drunk. She said that she had to ask him twice to move. Then, she turned her attention to me. She asked me three times if I was a diabetic and I finally responded, " Yes". They tested my blood sugar and it was a SIXTEEN. For anyone who is a diabetic, you know what that means. I was in shock. This is when they fed me glucose intravenously through a tube and needle and my head started bobbing around ten minutes later. They had already put me on a gurney and I was in the ambulance when I became conscious and was coherent enough to talk.

They asked me if I wanted to go to the hospital and I don't remember saying YES but I must have. We were on our way to the hospital. I would have rejected this offer from them, had I been more aware of what had just happened. Sofia was able to coerce me into going to the hospital for a more complete examination. To make my day EVEN worse, they tried to call my roommate and no one was home when they called. Everyone was at work. Ultimately, they called my boss at work to come and stay with me while I was there.

I was going through a divorce and didn't have the appropriate contact information with me for the hospital to call. I was soo embarrassed when I finally came around and saw that I was in a bed in a hospital room. My boss was sitting in a chair next to my bed. She was about seven months pregnant and the hospital was not on her way home. I didn't understand why the hospital had called her. They told me that this was the only phone number that they could find in my red wallet. So, I took a deep breath to relieve some tension to begin an intelligent conversation with her.

She was wearing a long, beautiful chocolate brown dress and a colorful scarf around her neck. She had on some lovely gold earrings too. She was tired and looked like she could go to sleep sitting in that chair. I told her how much I appreciated her being there and apologized for the misunderstanding. I wasn't conscious when the hospital made the call to her. Our conversation was short but had said volumes by the time I got back to work.

She was very nice and said that she was looking forward to her baby coming. She was losing sleep because she couldn't get comfortable enough at night to sleep. Her stomach was huge and it was difficult to find a comfortable position to snooze in. I told her that I was sorry that they called her, again. I told her that I was fine and was getting ready to leave the hospital as soon as they would let me go. I told her that the hospital had called Natalie and she was coming to get me. They had called my apartment but Natalie wasn't home. They left a message for her to call me at the hospital as soon as she received their message. I was awaiting her call. I wanted my boss to leave the hospital as soon as possible. Natalie had called and spoken with the nurse at the desk while my boss was there. She told the nurse that she was en-route to the hospital.

According to my boss, the cart of mail would be waiting for me when I got there tomorrow on top of the mail that they would be getting that morning. She didn't make it seem like this was significant but it must have been. I lost that job a week thereafter. Obviously, because of the stress, this job wasn't flexible enough to meet my needs. I have remembered and never tried to re-apply for a newspaper position like that one since then. Secretarial responsibilities seemed overwhelming to me because one such task was all that I could accomplish. I was fine, as long as I wasn't hypoglycemic. I had seen favoritism in this particular job toward military employees. I made a mental note to myself to be aware of it when re-applying for another position.

About an hour after my boss left the hospital, my roommate and her daughter arrived. I was discharged and they took me home. Natalie had to retrieve Karen from her boyfriend's place first before they came to get me. It was dinnertime by then and I wanted to go home to relax. The day had not gone well and tomorrow looked like it would be even worse, as far as work was concerned. However, we needed to go to the Chargers gas station where my car had been parked by the policeman before heading home.

I was beyond exhaustion and decided to let Karen drive my car home when we arrived at the gas station. She had her learner's permit and seemed to be a very good driver. Natalie followed behind us in their car. Traffic was horrendous because of the approaching dinnertime hour and people were departing from work. We decided to take a leisurely drive home. Karen was on her way to getting her driver's license and we got home without incidence.

When I walked into our apartment, I collapsed on the sofa directly next to our front door. I remember thinking when I sat down, why didn't I notice that I had driven all the way up to the 7-11 which was five blocks past our apartment complex, and just turn around somewhere? I remember thinking this when I saw that convenience store but I didn't know what to do next. While making that drive down Virginia Beach Blvd, I remember passing the Virginia Beach Public Library first and then, I passed that particular 7 – 11 store. Then Pembroke Mall and things started getting fuzzier after Pembroke. All of these places were on my right as I was driving down Virginia Beach Blvd. I had a slight head ache and my brain was muddled. It was very obvious by then that comprehensive reasoning which I've suspected for more than twenty years isn't possible during hypoglycemia. Also, I am not usually driving when my hypoglycemia is critically low. If I am not driving or working, some kind of a break is mandatory at that time.

Because hypoglycemia isn't a scheduled event at work, I navigate around it and keep Tic Tacs in my pocket for emergencies. It had been a long day for me and we were getting ready to sit down for dinner. My roommate, Natalie, and her daughter, Karen, were busy making dinner and I got up to set the table. They both suggested at the same time that I sit on the sofa until their dinner was ready. I didn't put up a fight and waited about ten more minutes. My blood sugar was normal but a liquid lunch doesn't stay with me for very long. It will raise my blood sugar quickly but there is nothing in my stomach to sustain it. I was starved.

This event helped to establish my priorities during the final stages of my divorce. This was the first time in my entire life that I had ever passed out in the driver's seat from diabetic hypoglycemia since my diagnosis twenty-five years ago. This was the first incident that year and there were two more to come, in that year. This showed me how much stress could affect my life. There was no other logical reason for why this happened. From that day forward, I knew that going through another divorce could never happen to me again. This meant that when I do remarry, it will be forever. Thus far, it's been a very lonely existence and I keep a tighter check on my blood sugars to avoid these instances. Marriage may happen for me again and it may not. I found that my own diabetic needs will remain to take precedence over my husband's needs sometimes. The future isn't as clear as a sunny day but some aspects can never change.

Chapter 16: Ah, Belligerence

Today was a day of errands all over this fun, beachfront city of Virginia Beach and it's accompanying cities and counties. This was a day that promised to be beautiful. It was the end of August and my air-conditioning was fixed in my car a month ago. It was still hot but not as bad as it was in late May. Especially since I had air-conditioning. I had the day off from an encyclopedia company job where I was working and needed it badly. I needed to do grocery shopping and stop by the pharmacy and the bank. I was even going to apply for another job so I dressed to do that first. Then, I could go home and change afterward. I left the apartment feeling ready for my interview. I was going in for another retail position at the mall and felt that the skirt and blouse I was wearing would make a nice impression on my prospective employer.

I had eaten half of a sandwich, feeling that this would be enough for me to make it through the interview and then, I could get something to eat in the mall. It seemed like it could be fun. Natalie and Karen had to work that day. I planned to go to the mall alone and enjoy it for a few hours before going home to change into something more comfortable to complete my daily errands. When I got into my car, I felt fine. I was a little nervous which always happens when I am going to an interview. This one seemed like a shoe-in with all the retail experience I had. I didn't think it would take more than an hour for the interview itself. What actually happened was unexpected.

When I got into my white geoprizm car to leave the apartment for that interview, I was feeling great. I knew that I would get this job. The interview was a formality. Leaving the apartment parking lot, I needed to make a right at that light to get to the mall. For some reason, I went straight across the street. There was a car in the right- hand lane and I didn't want to collide with it. Driving around that block and making a left onto Virginia Beach Blvd seemed like a better detour. But when I got across the street, my mind started going haywire.

I didn't recognize that street but knew that it was almost directly across the street from my apartment. I was having problems getting my bearings. All of a sudden, a police car came up behind me and its blue lights were flashing. I made a right hand turn to get out of traffic and pulled my car over to the curb. I couldn't imagine why he would be pulling me over and hoped that he was after the red sports car that was in front of me. No such luck.

I sat in my car and started feeling dizzy. My car window was down so I thought that maybe it was the heat. Heat can cause a myriad of symptoms for a diabetic, even contributing to low blood sugar. I was in a hurry to get this interview overwith to enjoy the rest of my day. Anyway, the policeman came around and said something to me and I don't remember what it was. I think he asked me to show him my driver's license.

I started to take it out of my bright red wallet. He could see that I was having trouble doing that and offered to help me. I handed it to him and he took my driver's license out of the wallet and returned my wallet to me. Then, he went back to his car and got on his car phone and checked my credentials. I was still using my red wallet and all of my old information was stagnating in that wallet. When he came back, he asked me if he could sit next to me in my car and I agreed. I guess he realized when he checked me out that I was not a threat to him in any way.

He said that I had weaved in front of another car and my memory was so fuzzy that I didn't remember doing that. I told him that I was on my way to an interview and that they make me a little anxious sometimes. He said that he could understand that. I still don't think this is why he pulled me over but I couldn't visualize what had happened and wasn't going to argue about it with a policeman. Then he asked me if there was someone he could call to help me. I was getting more uneasy because it was getting later than I had planned for this interview. Punctuality, even when planning and executing my errands, was ingrained in me during my education about diabetes in 1968.

The stress of being pulled over by a policeman and being late for an interview was grating at my nerves and bringing down my blood sugar. I didn't have my own cell phone so he needed to use his to call Natalie. I was getting antsy at this point and said something that I shouldn't have to the policeman that he later said was BELLIGERENT. I don't remember exactly what I said but I think I was trying to discourage him from calling Natalie. I apologized repeatedly.

He had called Natalie and this is exactly what she said; "FEED HER. She'll be fine." There was a 7-11 convenience store directly across the street from where we were parked and he went over and got me some saltine crackers and orange juice. When I became more like myself a few minutes later, he told me what had happened and this is when I sincerely apologized for being belligerent. My problem in these situations is that I think that I know more about what to do than most people and I try to articulate this. However, my mind is deficient of sugar and my thinking is impaired. It's impossible for my mouth to articulate anything correctly and I get tongue-tied. I tend to sound belligerent when this happens. I get frustrated because I know my words are not being said the way that I mean them. This is when people who know me tell me to shut up and eat.

Before he got out of my car, I thanked him for helping me. He instructed me to be more careful in the future, as my driver's license could be suspended next time. We had not tested my blood sugar at that time but an educated guess would imply that it was in the mid-20s. I promised him that I would do just that. However, at that moment, I still had an interview to attend and had to change my afternoon plans slightly. I stopped at a Subway and got my usual 6-inch tuna on wheat bread with onions, lettuce, and tomatoes before my interview. I didn't want to become hypoglycemic again and not get this job. It was a job at a department store and I knew I would get it. The most trying part of the interview would be negotiating for descent hours and an inflated salary. After lunch, I went to that department store and asked if they were hiring.

This was a store that wasn't open to the public yet and I had seen signs stating that they were hiring. They handed me a paper application that I filled out from memory. When I went in for that interview, I was hired on the spot. I met and liked my prospective manager and was informed that I would be working in women's clothing. They were so impressed with me that I was hired to sell interview suits and essential accessories like hosiery and shoes and purses. In other words, I would be their personal shopper. The rest of my day was pretty uneventful which was the way I liked it.

After the interview, I went home to change. It was around 4:00-ish by then and I still had time to go to the pharmacy and the grocery store and get home in time for dinner. I didn't feel that I had time to go to the bank and decided to wait until tomorrow to complete that errand. Natalie and I talked at length when she got home from work that day. I was turning out to be more of a responsibility than Karen, her seventeen year old daughter. Natalie was fortunate that Karen was never sick a day in her life.

I promised her that I would try to get things in order quickly and told her that I had gotten hired for a department store job just that day. She knew I hated selling those encyclopedias by phone and needed to get away from there. She was happy to hear about my new job and she wanted to take me out for dinner that night. I was free and gingerly accepted. Karen was with her girlfriends that evening and wouldn't be home until late.

That day turned out to be great, even if I had been pulled over by a policeman. I had still gotten most of my errands done and landing that job was the icing on the cake. We went to a family restaurant which was a nice way to celebrate. Both of us worked frequently and rarely had any time to spend together. Her daughter was still in high school. It wasn't the best job I could have gotten but it was definitely a step in the right direction for me. I was hoping that the next few months would settle down.

I realized two months after I started my new job that I would need more money from the job to help pay the rent and electric bills and still have money left over for groceries and my medicine. At that moment, I was going to enjoy my new position and the somewhat inflated salary that it could provide. It was something more than the commission I was making from selling an occasional encyclopedia by phone.

Selling encyclopedias was one of those jobs where they paid you less than minimum wage because you received a commission on your sales. I had a conscience which made selling encyclopedias impossible for me. The thought of using a phone to conduct business was more appealing than face-to-face transactions. But not to sell encyclopedias.

After dinner, we decided to stop by the grocery store and get some wine coolers. When we got home, it was 9:30 and Karen wasn't due home for another few hours. Natalie was getting tired and I was oozing euphoria from snagging that job. Natalie was always tired by 9:00-ish and ready for bed. I was still going until at least 11:00P. Natalie was envious of all the energy I possessed. Having to be aware of how you feel on a 24/7 basis makes your day never-ending and makes your life challenging. From that day forward, I vowed to keep a more fervent check on my blood sugars to avoid incidences in the driver's seat. One thing about me; I never back down from a challenge.

One challenge I wanted to face and resolve without further complications was my job stability. This was my answer to making my life less complicated and, hopefully, bypassing another embarrassing situation involving a policeman and my belligerence. Karen had told me, once, that her mother had a difficult time paying bills and was known for forgetting to pay one occasionally. I was constantly reminding her of when our rent was due so that she would be prepared when that day came. Same thing with the phone bill. I knew that if I could make more money, it would help us out with our expenses. She and Karen had left a very abusive situation with her husband and she was in fear for their lives, when I first met her. She, later found out that her husband had moved his girlfriend into their house with him, according to Natalie's mother who lived nearby him in New Jersey somewhere.

Natalie had her own problems and her heart was made of pure gold. She was one of the sweetest and most honest people I had ever met in my life. We clicked the first day I met her at work and became friends immediately. We spoke briefly and I mentioned that I was looking for a new place to live. She told me to give her a call after work that day and that is what I did. She told me about a room that was available where she and Karen were staying and suggested that I look into it. We ended up rooming together and the rest is history.

The challenge of finding a lucrative job was on and I was determined to find one. There had to be a financially sound job somewhere that was perfect for me. Unbeknownst to me, I would face one more episode in the next few months. My belligerent episode with the policeman was a close call and could have been much worse. Interviews are stressful for me and my nerves could cause my blood glucose to drop irrepressibly. Controlling my blood glucose became more important than finding a lucrative job. I was losing control of my blood sugars and hence, my diabetes. Life's demands were just beyond my reach at that time and finding that ideal job was going to have to take a backseat to my hypoglycemia. I could wind up unconscious or dead somewhere before ever discovering that job. Belligerence doesn't seem that bad when compared to hypoglycemic shock which could cost me my life, next time.

Chapter17: Control

It was obvious after my second month at the department store job that I would need more money. I had been looking for something else, when a day off finally presented itself. The only thing it looked like I was qualified to do was to ring a cash register. After that second month, my bills were getting much bigger than my paycheck and Natalie was starting to spot me money for groceries, sometimes. My credit card balance was getting outrageous. My ex-husband was paying my monthly insurance bill and was due to end paying it in three more months. However, my medicine kept getting more expensive.

I needed insulin to survive and could only afford to use one test strip per day. After that, I had to judge my blood sugar level on how I felt physically. This is the worst way to determine what it was but when you don't have enough money to afford the insulin AND test strips, you forego the test strips. Monetary issues were becoming more and more of a problem.

I had called about a job working at an advertising agency. It was something that I had never done before but felt that my education would be more than adequate for this position. However, because most of my job history revolved around customer service skills, I hadn't included my job with the newspaper company. I had dealt primarily with the mail but it included other duties in the publishing department as well. Therefore, they didn't see where I had any experience creating advertisements because of what happened with my pregnant boss. Somehow, I was able to talk them into granting me an interview one week after my phone conversation with their agent, Veronica.

Her and I liked each other immediately when we spoke on the phone. She scheduled an interview for me exactly one week after our phone conversation. She said that she would be conducting the interview. She mentioned the formalities of this position and had to be professional. I understood that and was very optimistic about this opportunity. It was an 8A – 5P job with every other weekend off. I couldn't believe it. The salary was wonderful too and much more than what I was making at the department store. I knew I could do it. I just needed to convince her of that.

When my interview day arrived, I felt ready. But I was extremely nervous. My experience at the newspaper wasn't extensive but it could help me in the interview. It couldn't be included on my application. My college education included mostly English classes which was why I received a Liberal Arts degree as my first collegiate accomplishment. College had been fairly difficult for me in mathematical concepts but English was my strongest skill. I took all of the required Mathematical courses, up to Precalculus and passed them. Calculus was too difficult for me. Sometimes, I wasn't sure if it was the teacher or me that didn't understand what was being taught.

The English classes that I had taken extended into foreign languages. They included French and Spanish courses and I still remember some of it, twenty years later. The saying of " Use it or lose it" applies to me here because it wasn't utilized at all after I took it. However, my English classes were all advanced. I knew I could handle an advertising position, if given the chance. When I got to the agency, I walked inside the building and spoke to their receptionist. At that moment, I was informed that someone else would be conducting my interview.

My stomach dropped, all of a sudden. I felt a little weak and sat down in a chair in their waiting room. I started sweating almost instantaneously. Ten minutes after I spoke to the receptionist, another employee, Kelly, came over to me and introduced herself. She apologized for Veronica's absence and said that an emergency had arisen and she had to leave the office. Kelly ushered me into one of their interview rooms and gave me an application to complete. She said that George would be conducting my interview and that he would be with me shortly.

I completed my application and was getting more nervous as I waited for George. When he came into the room, he was dressed in a nice, gray suit. He had dark brown hair, blue eyes, and a medium build. He was about my age at that time. He looked professional and like he had been there for awhile. We introduced ourselves to each other and he asked me for my application. I handed it to him and he looked it over to make sure it was complete with my address and phone number information. He asked me a few direct questions pertaining to what I would be doing, or so I thought. We liked each other when we met, almost like Veronica and myself.

He was nice and polite and seemed to know what he was talking about. I muddled my way through one or two of his questions and then he continued with some more specific questions. When the interview was over, I asked him when he thought I might be hearing from them? He said to give them a call in a week if I didn't hear anything. I asked him if he had a business card and he said they had run out of them. Kelly had just ordered them that day. He seemed sincerely interested in me for this position. I remained fairly optimistic but still felt a little skeptical, since I wasn't able to interview with Veronica.

As I left their office, I was in a daze. I wasn't completely sure how I had felt about the interview and wanted to call and speak with Veronica. I walked down the steps and sat in my car for a few minutes before I left their property. I pulled out of their parking and into traffic and made a left turn. My mind was still dazed about what had just transpired and I started yawning. I assumed that the distractions in that interview created my fatigue. My mind wasn't clear when I was pulling out of their parking lot. It appeared that I was not heading toward home. I was headed into Norfolk and not Virginia Beach, where I lived.

Traffic was getting heavier and I couldn't stop yawning. I didn't know where to turn to get headed back the other way. I was having trouble focusing on the road and looked for a convenience store where I could stop for a minute and ask for directions. Eventually, I saw my chance to make a right turn, hoping that I could make right turns around the block and make that final left turn to get myself headed toward home. Instead, I ran into one-way streets and kept getting turned around. Finally, I came to a stop sign and lost consciousness for a few seconds behind the wheel. Hypoglycemia was setting in for the third time this year. .

I was out of my city on this one and had to trust the Norfolk Emergency Services Team. This time, their response team was almost directly behind me. The ambulance was about three blocks behind me when this happened. The traffic was mounting behind me and it was taking me too long to make that turn. People were getting into the other lane to get around me. One of the technicians spotted me and realized that something was wrong. They flashed their lights and pulled up behind me. I was conscious enough for them to give me something to eat without needing to inject me with glucose.

At that moment, they gave me a glucose tablet which I ate without any complaints. We sat there approximately fifteen minutes before they would allow me to leave. They asked me if I wanted to go to the hospital, like they always do, and I declined. No policeman arrived during that time and I was relieved. I was ready to make a right turn and drive away. They turned off their lights when I pulled around that corner to make that turn.

Shortly after that moment, I had myself going in the right direction to get home. I was drained from the interview and wanted to get home to change. I needed to relax. Once I got home, I fixed myself a sandwich and sat down to watch some TV. I started thinking about how else I could keep the lows from happening to me. I had candy with me and seemed to forget about it when the time came. I was drained, as usual, after an interview and an episode. The wheels were turning in my mind and seemed to have no where to stop.

I continued to rest for an hour and watched the television. When my mind started focusing on the days events. I decided that it was a good thing I was in Norfolk and not in Virginia Beach this time. They might have suspended my driver's license otherwise. I really wanted this job and called them back six days later. They had already filled the position when I called. There was another company that I wanted to work for and had been checking the newspaper for any open positions. I gave this opportunity three weeks to materialize and it never did.

After having so many hypoglycemic episodes in one year, I decided that I needed to take control over my life and reduce these incidences. Because I was afraid that my license would be suspended by the city of Virginia Beach, I started making arrangements to move back to Richmond with my immediate family; mom, brother, and sister. I had grown to love Virginia Beach and truly didn't want to move. This was the only way that I could determine how I could have better control over my blood glucose and my life. The next few months would decide my future.

Traffic lights and stop signs have proven to make me STOP, especially in my subconscious mind. Whoever created the idea to use them on our roadways deserves recognition and I want to thank them too. There are a myriad of people who I think should be recognized and those who have produced these tools to keep traffic flowing smoothly are included. My elementary school teachers were the ones who introduced me to STOP lights and STOP signs. They are the first pioneers to enter this idea into my mind, when my mind was a sponge and remembered everything. Traffic lights and stop signs were the barriers that brought order to my life during these episodes. Soon, I would have the resolution to this problem under control and these incidences would become minimal. I could see my resolve coming and the light at the end of that tunnel was getting brighter.

Chapter 18: Resolution

I regret that I had to pass out behind the wheel three times before I finally realized what I needed to do. I was trying so hard to make it on my own that my diabetes was suffering for it. The architect that I had been dating was my savior during this part of my life. He was financially set and seemed to have his life in order. He invited me over on weekends to be with him. He seemed to be looking for someone to be with just like myself.

He wasn't perfect but was much better than what I had with my ex-husband. His name was Kevin. His job was more stable and didn't involve any type of weaponry to perform it. He worked for the Navy also but in a more subtle capacity. Feminine traits or qualities seemed to come naturally to him because he spent his time studying the human body in an artistic format to acquire his Masters Degree in Architecture. He wasn't a homosexual in any way and wasn't really aggressive about anything which was a nice change for me. This appealed to me after what I'd been through.

Kevin was brilliant. Physically, he had dark brown hair and eyes and was my height and small, like myself. He was someone who wanted to do things like go to the beach to ride bikes instead of sitting in front of the television all weekend. He had every weekend off. I found out later that he was not that happy with his job because of the people he worked with. He made six phone calls to different employers around the United States that last month that we were together. He was hired by the fifth employer in San Francisco, California. He eventually married an oriental lady whom I'd never met, bought a house in San Francisco, and moved. This happened within six months after we stopped dating. Apparently, he had been looking around for a new job for more than a year..

I had moved back to my hometown, one hour away, with my family by then and was only down there on weekends. He finally told me that he had met someone else and I was crushed. I was just getting comfortable with him when this happened and wanted to continue our relationship. I might have discussed a more permanent relationship with him if he had not already made other arrangements. My divorce had devastated me and this blind-sided me, as I hadn't wanted to make him feel cornered by moving in with him, instead of moving home with my family. I didn't think it was fair to him.

I remember one weekend that we went to a tennis court in his neighborhood to play tennis. We had been playing for about twenty minutes when I decided to start playing more competitively. In doing so, I went for a ball that barely made it over the net and I was in the backcourt area. I broke my right big toe or great toe, as the doctor explained to me thereafter, when I attempted to hit that ball back over the net. I was in a full-stride step and broke it when I went to take my next step toward that ball. I didn't know that it was broken and felt that it was ok when I first did it. But then, the limping started. He talked me into ending the game early and asked me if I was sure that I was ok. He offered to take me to the hospital and I declined. After I got to Natalie's place later that evening, I took off my shoe and procured a better look at it.

We had taken my shoe off earlier to look at it which was why he wanted to take me to the hospital. He had an icepack that we put on it while I was at his house. I took myself to the hospital shortly after I arrived at Natalie's and found that I had indeed broken that toe. They gave me a lovely blue shoe to wear and taped the big toe and the toe next to it together. They told me to watch the dressing and make sure that it stayed clean. They also said that it would be better to elevate the toe whenever possible. I was just upset that I had broken my toe. It was sore to the touch. Man, what a pain.

While I stayed with him, I examined his Sunday paper to see what job opportunities were available. I was so desperate to find something that I would have taken anything that would have kept me in that area. I liked living at the beach and would still be there today if I'd been able to find lucrative employment. But it never happened for me. So, I came back to my hometown and have been living with my family since then. I had tried living with roommates and I liked having roommates. There had been some unfortunate problems with the boyfriends of my roommates that I couldn't tolerate. One had been stealing my insulin syringes while everyone else was at work. He would have excuses to hang around our apartment frequently by himself when we needed groceries or needed to run other errands. He started hanging around our place when we were at work, all day. I trusted my roommate, not her boyfriends.

I had another roommate who ruined my clothes because she smoked like a smokestack. I asked her if she was a chain smoker and she swore that she wasn't. We only lived together for two months and that is why she is not mentioned in my book, except for here. I talked to numerous other possible roommates and none of them were viable. I had tried everything possible so that I could stay. It was evident that I needed help particularly financially and couldn't live somewhere without having a roommate. However, I couldn't have lived with different roommates all my life because all of the moving and adjusting to different people was too stressful for me.

Therefore, I moved back to my hometown and stayed with my family for the resolution to my financial woes. They are financially secure and this takes that one overwhelming burden off of me. If I didn't have the diabetes to contend with, the rest of my life would be more manageable. I wouldn't have to worry about passing out at work and losing my job. Being categorized as normal is another dream that I was never able to see come true. The cure may become available during my lifetime and it may not. However, in the meantime, I have a loving and supporting family that will always be there when I need them. Others are unable to understand me as well as they do and my family, especially my mother, has been rock solid in my more trying times.

There were two other episodes that year where I passed out behind the wheel from hypoglycemia. I did not go into any details about them because they were sketchy. During my second episode, I was in a residential area around 6:00PM. Wal-mart was my destination when I was driving home after work. I drove down this paved road for five miles before I reached a stop light. This time, I put my foot on the brake at that stop light, just like the previous time. It was a one-lane road with a left-hand turn lane at that light. Apparently, the lady behind me called 9-1-1. When the EMT's arrived, I came around within fifteen minutes. My blood glucose was nineteen points when checked. I did not go to a hospital afterward and did not lose this job because of it. This was the scenario that I strove for after that first incident on VA Beach Blvd.

The third time it happened, there is only one thing that I remember being told. When my blood glucose was checked this time, it was unreadable. NO number came up on their glucometer. This was when my diabetes altered the course of my future. It was obvious that I couldn't deal with the diabetes COMPLETELY on my own. As mentioned earlier, this event caused me to have an epiphany and it was to move home with my immediate family.

People to Remember

Chapter 19: Strangers

Strangers is a story about a not so dangerously low blood sugar, to begin with. I was living in my hometown during this occurrence. I was on my way home from work one evening and decided to stop in a gas station/ convenience store for something. I wasn't exactly sure what so I let my mind wander as I walked through the small, enclosed store. I wound up buying some blueberry mini muffins, M&Ms, Tic Tacs, and a diet coke. As I walked to the counter to pay, I started talking to the few people in the store. That particular night, there was a state policeman in attendance. I was talking to everyone and laughing mindlessly. I didn't pay too much attention to what was going on because I was close to my house and planned to go directly home after I paid for my purchases.

As I went to leave the convenience store, I started to open the M&Ms and decided not to. My house was three blocks from this store. But when I pulled out into traffic, a car went speeding by me in the left- hand lane and distracted me. I missed the right-hand turn that would have taken me into my subdivision and straight to my house. So, I kept driving. I was headed toward an area that I didn't know very well and it was getting dark. My stress level was starting to climb. Then, I checked both mirrors to see if I could make a U-turn to get me going back toward my house. When I looked in my mirrors, I saw headlights but they were far enough away for me to make that turn. I decided that I would make that turn and when I made that turn, a car magically appeared and pulled off the road across the street from me and stopped. So, I stopped.

I was already drinking my soda at that time and opened my muffins to eat them before this guy got out of his car. There were three people in his car, including him. I was alone in mine. All three of them were olive skinned with black hair and brown eyes. They spoke Spanish and I believe that they were Mexican. The man driving was in his late twenties and the young lady with him looked twenty-five. There was a little girl around seven years old in the car with them also. When their car had stopped, the guy and the adult lady got out of the car immediately. I remained seated in my car. The little girl had been sitting behind the young lady and got out of the car to walk toward him. He yelled at her and told her to get back in the car. I just sat in my car and kept eating.

The problem with this particular situation was that he was drunk. He had called the police and said that there was a drunk driver on the side of the road who he identified as me. I didn't have a cell phone at that time so I had to wait and just kept eating. I was so happy to see the police officer when he arrived. He stood and talked with the other guy for a few minutes before he came over to talk to me.

The officer came over to my car and asked me if I was ok. I told him that I felt a little low earlier but that I had been eating in my car while we were waiting for him and he let me go. Apparently, he had whatever he needed from me. The other guy was still there when I pulled away. This was the only incident where I believe that the stranger was in an altered state of mind and made the wrong call for him and the right call for me. He had called the police which was the best thing that could have happened in this situation for me and his passengers. They received help before things got out of hand.

It was a blessing for me that I stopped at that convenience store when I did, even if I was less then a mile from my house. And why *DID* I stop at that convenience store when I was so close to my house? My subconscious mind must have been watching out for me, as it has done so many times before. It's times like these when I have learned to trust my instincts. Strangers have been my saviors in the past so maybe it was my turn to return the favor this time.

Chapter 20: Friends

Friends have been a very important part of my life. Even people who started out as strangers to me initially became my friends, once I got to know them better. It's tough being a role model to other diabetics who are friends of mine. Most of them are type two diabetics and dealing with my diabetes is obviously different than theirs. There is a very fine line between the two types but the line exists. I've had many friends in my lifetime who were diabetics and we all seemed to scatter in opposite directions. I would be strongly interested in Volleyball and they were strongly interested in Swimming. This was the type of thing that separated us, most of the time. We needed to be physically active and if we chose different activities to participate in, especially in high school, then we had to learn how to delegate our time productively.

When I was a child, I was devastated when I lost a friend. As I grew older, I realized that friends weren't always going to be around when compared to family. However, since I've never had a family of my own, a few of my closest friends have become almost like family to me. Neither family nor friends may be around when hypoglycemic shock happens, depending on where you are at the time of an episode. Usually, explaining situations that have happened to me to a friend is MUCH different than explaining it to my mom, brothers, and sister.

I've told more than one friend of mine at different times in my life that if I'm not passed out somewhere, I'm happy. My friend or friends, not realizing that I meant from hypoglycemia and not from being drunk, couldn't quite understand my meaning behind that statement. I always knew what I meant whether we were close friends, or not. I also don't elaborate too much because it seems to confuse friends and other people, unless they know I am talking about my diabetes. AND most of the time, I am.

I enjoy every minute I spend with a friend. I've made many friends in the past who have been there when I really needed them, without judging me. My family is wonderful and my mom always thinks she knows what is best for me. This is typical behavior for moms. My dad passed away many years ago but he seemed less judgmental than my mom. But, I couldn't love anyone more than I've loved my mom and my dad. When I was a child, most children couldn't understand what was happening to me and why I wasn't acting right.

In the early stages of my diagnosis, I didn't understand it that well myself. But I kept trying to explain what was going on and it seemed to make things worse. So, I stopped explaining it and let my friends see what would happen. I was always back to my old self afterward and we would go back to our usual activities like homework, television, or playing in the back yard. It was a learning process for me and I learned that it was required for me to have something around to eat all the time. All foods have some sugar in their contents and sugar is what is in demand when I am low.

I learned how to control them. They would surprise me sometimes, especially when I was hectically busy. My friends and family can tell when I am not myself and feeling like this. Just ask my mom. However, the symptoms I display when getting low are often confused with normal, every day behaviors like yawning.

My drastic mood swings have been viewed by others as PMS or Premenstrual Syndrome.

Do you recognize any of the symptoms listed? Blurred vision, dizziness or light-headedness, fatigue, head-ache, laughing, nausea, nervousness, slurred speech, sweating profusely, weakness, and yawning. If you are a diabetic, at least three of these symptoms will look familiar. Having close friends can be a comfort when hypoglycemia occurs. My life would be much more complicated without their friendship. Friends are as precious and delicate to me as a flower. Flowers require food and water to survive but continuous weeding and mulching makes the plant bulb flower and blossom, like a friendship.

Situations to Control

Chapter 21: Stress

Stress plays a big role in how my day flows. It is a part of my life and everyone else's continuously. Stress management is essential for helping to keep me in balance, mentally and physically. This is why things that used to bother me don't anymore. It isn't a good feeling to wake up in an ambulance and then, your mind starts to put together some of what happened beforehand. It takes me at least twenty minutes after I awaken to start thinking clearly The Emergency Medical Technician (EMT) or someone involved in this process has to tell me what happened while I was unconscious.

One night, I was in a club in the beach area after my divorce with my friend and roommate Natalie. I was talking to another friend of mine and I saw a face that looked familiar to me. She came up to me and started talking to me. The first thing she said to me was; "You don't remember me, do you?" I said, " Not exactly but your face looks familiar." The light bulb went on above my head when I saw her. She said, "We've picked you up a few times when you were hypoglycemic." Initially, I was slightly embarrassed and then, I began to remember how nice she was. I knew what she meant as soon as she made that statement. She said, "My name is Sofia and I know you as Britta McGeorge." She had definitely seen my driver's license. I said, "It's nice to meet you, Sofia. My friend's call me Brit. Don't take this the wrong way but we've got to stop meeting like this." She started to say something else when Natalie broke in and said that she needed to leave. She had heard Natalie. We smiled and waved good-bye to each other.

When Natalie and I arrived home that night, I thought about that short conversation. I had wanted to talk with her for a few more minutes but something was up with Natalie and I was always the designated driver. I haven't been much of a drinker since college. It was apparent to me, after passing out in the driver's seat three times due to hypoglycemia, that I needed to make my life less stressful. My financial situation at the beach was making life unmanageable. All of the repercussions from my divorce came crashing down on me and I finally decided that I would have to go back to my hometown. If the police decided to snatch my license and I had NO relatives around to help taxi me back and forth to work, then I would be in a world of trouble.

I came to my hometown and found a descent paying job within a month after arriving in town. I didn't have any other choice after living at the beach for the last five years because of the diabetes. I had spent two and one-half years completely on my own, living in that area with a roommate. My husband and I were together for one and one- half years and were legally separated for one year.

My experiences with stress at the beach were life - altering. I realized that stress could immobilize me and make it impossible for me to work and make enough money to pay for what I needed to survive. The three main ones: Insulin, Food, and Shelter. I checked into some kind of government assistance program and was not eligible for food stamps. I was told that I made $1.00 too much per month to qualify for it. This increased my fears and showed me how close to poverty I really was. I didn't feel that the government was going to pay for my medical bills and believed that I had a better chance of getting food stamps. I was so upset when I left there, that day, and felt completely helpless. I didn't know what to do or where to turn. If the government wasn't going to help me, then who was? I knew I couldn't deal with the diabetes alone.

I have learned that I need to keep an extremely tight rein on stress. It can obscure my priorities in life to the point of death, without me being aware of it. I haven't tried licking envelopes and sending them out in the mail because of what I heard from my sister. She says that it's a lot of work for only pennies a day. So, I keep getting jobs and hoping that just one of them will work out for me. Now, I am trained to work in a pharmacy but am currently focusing on customer service skills until our grocery store upgrades to a better store that will include a pharmacy. If my particular store does not include a pharmacy, it will be possible for me to transfer to one that does. I don't want to lose any tenure that I've already obtained from this position at my store. With the state of our current economy, I don't want to take that chance.

There have been times when I feel that I have lived for my diabetes and not with it. I am living for the diabetes anytime stress enters my life. Stress can cause my blood glucose to decrease rapidly, as well as waiting too long to eat. I always try to control stress before my blood sugar plummets downward. Non- diabetics experience the same sort of drop in blood glucose around mealtimes or when they are excessively hungry. The main difference is that if you have diabetes, your body can't keep your blood sugar from dropping critically low. If you are a non-diabetic, your body maintains a normal blood glucose level naturally or without the aid of prescription medicine. Hunger can affect a non-diabetic and a diabetic in similar ways. Except with diabetes, you have to address it when it happens to prevent hypoglycemia. This hasn't happened to me often but when it does, it's quite memorable. I try to do things to prevent it from occurring.

People say that they have seen me eating in my car frequently. It's quiet there and I can eat something without any interruptions. Hypoglycemia reconciled me of cleanliness over godliness. I'd rather be conscious before clean and safe rather than sorry. I can always stop somewhere and vacuum my car out when I'm conscious. If you aren't a diabetic, count your blessings. Diabetes is a 24/7 job, with no beginning and no end. Particularly if you are a type one diabetic.

Chapter 22: Work

To me, diabetes isn't as complicated as work. Work doesn't coincide with diabetes and vice-versa. At work, it feels like I am in the military because they own me for however long I am there. My diabetes doesn't understand this. I can't leave my diabetes at home before I go to work. The two are inseparable. With diabetes, the rules are very straight-forward. There can be FEW to NO gray areas when dealing with diabetes. Trying to establish my professional life has been the most unsuccessful venture I've ever attempted, even marriage. With marriage, there was an end to it and with work, it will continue until my retirement at 62 – 65 years old. I seem to get caught up in customer service jobs that have no future. Things seem to be easier for me to deal with on my own.

I would call myself an eccentric intellectual with a twist. The twist being the diabetes. Working makes me feel productive and puts some money in my pocket for necessities like food and insulin. The diabetes is the one thing that keeps my life in focus. It takes precedence over work. I've tried to neglect or ignore it before and this has proven to be fatal, a few times. The diabetes establishes how my body functions and my body tells my mind what to do.

Even with years of college behind me, that perfect job continues to elude me. I remember my dad saying that money makes your life more comfortable. But with the demands of our current society, money is more of a need than it has ever been before. After my experiences with stress, the diabetes remains to dominate employment. The process of getting hired has become a job in itself because of the internet and using online applications to apply for a job. When applying for an online job, I can't be sure if or when which store will contact me for an interview or employment. Usually, there isn't a number listed where I can call them to inquire about the status of my application.

I am most comfortable sitting right here in front of my computer, doing what I am doing right now. Typing. Apparently, I'm not fast enough to be a data entry clerk and make more typos than is deemed necessary for me to be hired as a data entry clerk. I don't believe I make that many typos but somebody else does. I love the internet except when I get a phone call. I am on the old dial-up system because I don't want another piece of hardware connected to my tower. Therefore, when a phone call comes through, I get disconnected. I've learned to go online when I have the fewest phone calls coming into the house.

Most of the time, I end up in customer service positions which require me to operate a cash register. Using a cash register became easier when I started using automated registers. The register does all of the math for me, as long as I am entering the correct information. All prices are pre-set before you buy an item. When I scan the barcode, it will give me a price. That part of the math is done for me. It's when I go to collect the funds for it is when things can get a little shaky. I have to hand enter what a customer gives me in cash and if the register only reads 2.01 instead of 20.01, then I will add in 18.00 to make up that difference. I may not have pushed the zero key hard enough for the register to read it.

If I don't see that the register did this, then it might confuse me for a minute. I'm pretty good with these types of mistakes most of the time. I let the register do the work. I make sure that whoever I'm waiting on gets what they need. Product prices aren't truly negotiable because they are pre-set. However, prices change when we start a new sale so it's possible to negotiate the price with a manager. Most registers today are automated and all retail establishments have one to keep a track of sales and provide a customer with a legible receipt.

Before I had this job in the grocery store, I had another job in a department store. These jobs are available everywhere. I am definitely over-educated and way under-paid. Because of my college education in the field of social work, I am usually able to determine what someone is like relatively quickly. This hasn't helped me in interviews but has in numerous other situations. Young children can cause distractions for parents and other customers. This can create a domino effect and produce other distractions, even for me. We all have bad days so I usually reserve judgment until after I have helped them a few times. Most people are very nice and I like them, instantly, there is no doubt about this.

Working with someone and being friends with them are two completely separate relationships. People's behavior patterns are learned and situational. When you learn how someone behaves in a certain situation, they will mimic this behavior in a similar situation. These are the types of details that I remember about most people that I meet. I may not remember their name but I will remember their face. When I first meet someone, they may not be having one of their usual days and it is more difficult for me to establish a rapport with them. I might miss something the first few times we meet because of my job responsibilities. I get distracted, as they do sometimes, and we work together to ensure that they are getting what they need. I use one of the ten commandments when waiting on a customer: Do unto others as you would have them to do unto you.

This is why I am efficient in a retail environment. I have an abundance of experience in this field and have some pre-determined behaviors that I hope will not offend anyone. At least, not until I get to know them. If I don't know them, then I may not know much about their likes and dislikes. Sometimes, I can't remember everything. I am too busy to notice. I always say that I am not perfect but I sure try to be. People choose to buy certain products because of their personal beliefs and extenuating circumstances such as family members and their dietary needs and preferences.

For example, some people will buy organic products instead of the store brand ones. Maybe they didn't buy exactly what they wanted this time because it was too expensive or we were out of stock in that product. Whatever their reason for buying something, it is their choice to make whether I think it is a good one or not. Their wants and needs are different than mine and, as you already know, I have a few of my own needs. I try to make their shopping experience a pleasant one because I want them to come back. I know from my experiences with hypoglycemia that people are innately good and I try to extend the same courtesy to them.

I work very well in call center environments also. The hours are consistent and this is important to me, even moreso than the pay sometimes. For instance, I was working in a call center on a Saturday. It was a very laid back day and we weren't receiving that many calls, to my surprise. I was working an 8A – 6P shift that day and when 12:00 noon rolled around, I was feeling strange. I had eaten a big breakfast that morning and thought maybe I was running a little high. Highs make me feel lethargic. I had been having some problems keeping my attention on what the customers were saying but it was mostly; What balance do you have on my account? - And - What was the fee for? Obviously, I was working in a call center for a bank.

I started noticing my inability to focus on the conversations closer to 11:30 but by noon, I was feeling weak. My lunch break wasn't scheduled until 1:30. Since it was a Saturday, the floor manager was responsible for the second and third floor call centers. I couldn't locate him and he didn't notice that I hadn't answered a few of my calls. I don't know how many calls I missed, but I knew that I needed sugar and there was a Pepsi machine right behind me in their break room. I heard a beep in my headset but didn't respond to that call.

My purse was in the drawer of the desk where I was sitting on that Saturday. I started getting change out of my change purse for a drink. I didn't have any candy with me that day so the drink machine was going to have to do. I took my headphones off and laid them on the desk in front of me. I stood up and walked fifteen steps to that machine, with the change purse still in my hand. When I got to the machine, I put the money in and chose a Pepsi. I was lucky that it didn't eat any of my change. I got a nickel back and spent some time trying to get it out of that slot. My fingers are short and I always have trouble retrieving my change from those vending machine slots. I had started drinking that drink while I was fishing for my nickel. When I got back to my desk, I checked my watch to see when I could take my lunch break. Our floor manager had been on a call on the floor above me when I needed that break.

Anyway, I started feeling the effects of that sugar instantly. I am not sure how hypoglycemic I was but my educated guess would be a twenty or so mg/dL. Mg/dL means twenty milligrams per deciliter per drop of blood. That's the technical wording for what it means. Twenty to thirty points is what I usually would call it, to avoid the technicality. If you know your numbers, you know what that means. No one mentioned to the manager that I had left my desk for a drink. Had I been in my usual state of mind, I would have put my phone on stand-by to re-direct calls while I went for that drink. We weren't so busy that the phones needed to be monitored that close. No manager ever mentioned how many calls I missed that day so it must not have been too many.

The rest of my day was uneventful. I was a little tired afterward and started coming around pretty quickly after I had that Pepsi. I drank half of it and threw the rest away. Pepsi is extremely sweet to me which was what I needed at that time. I didn't look any different afterward and most people were not even aware that anything had happened to me. When I've tried to tell my co-workers what they should do before an incident like this happens, they have a tendency to panic. I usually say, "Dial 9-1-1 first and don't think about it. Just do it. They will ask you a few questions and will be there within ten minutes after you call." But for me, I'm the one who is unconscious for ten minutes after they arrive, or I'm conscious and in a fog. The only time it is truly obvious that I am hypoglycemic is when I am in my car. Then, it can't be missed because I'm blocking traffic. Hypoglycemia happens quickly and lasts ten to twenty minutes, depending on when I notice it.

Another part of dealing with hypoglycemia is knowing when it is a dangerous or not so dangerous low. It takes me some time to determine which it is because I have to consider what I've eaten, how much insulin I've taken, what my blood sugar was last, and the effect of these factors on me combined. I have to recognize my symptoms and what they mean. I caught this low in the call center before it got dangerous which was amazing because of all the distractions. There is more to working than answering those telephones. I felt tired but not so tired that I was lethargic or unable to keep my eyes open. Had I not caught it within that thirty minute timeframe, things could have turned out differently.

There is a major difference between these two stages of hypoglycemia; hypoglycemia – I am low but awake and hypoglycemic shock- I am low but unconscious. If I get so tired that my speech is slurred, then it's a possibility that it is a dangerous low. OR I could be excessively tired and yawning every few seconds. I wish that the doctor had a way to show me how to determine how low I can get BEFORE passing out. My idea is 15 points but it's unconfirmed.

There is a twenty-four hour blood glucose monitoring regiment that would require me to wear a patch on my stomach for twenty-four hours. I opted not to use it because I already have a cannula that stays in my stomach for four to five days continuously and this would be very uncomfortable for me. Also, it is not cheap and I can't afford to have both.

I need the insulin and can manage with the external blood glucose monitoring system that I've been using for forty years now. I have a pretty good idea of how low my blood sugar can get before it's dangerous but I wish there was something less constrictive, better, and much less expensive to use. Since there is no way of doing this yet, I will have to keep doing what I'm doing. I very rarely have dangerous lows but when I do, being prepared is fundamental.

I got a call from a representative about my insulin pump supplies recently. We talked at great length about my order. I had tried to call it in the day before and the supply order call center was not open when I called. So, I tried going online to do it and their computer system wouldn't accept my password. There was a link provided that said, Contact Us and I did. A woman called me about that order on the following day. She was nice and asked me some questions concerning my pump. When we concluded the call, she said, "It sounds like you have a system that works for you. You must be doing something right because you are still here to talk about it." I agreed with her and concluded the call. My 'system' is not written in stone but I am quick to notice subtle changes when my mind is alerted to them. This is how I survive.

I enjoy working because of the relationships I establish and the money. Of course, for me, work is even better when our machinery is working like it should be too. The only parts about work that I can't really control are the people and the schedule. I think this is true for most employees. We all work better with some people than others and some schedules, like day or night, are preferable over others. I've been involved in many middle management positions where the responsibilities were numerous. My mind usually handles this well but my body will take over, at some point.

As soon as I pass out from hypoglycemia, this causes a scene in the work environment. Things seem to snowball downhill from there. One manager is on the scene and tells another. Then, when the next schedule was posted, my hours decreased from 38 - 42 down to 30 – 35, and so on. Ultimately, I would either resign or quit that job and end up looking for employment elsewhere.

I learned that resigning from a job was better than quitting for money and reference purposes. I could find a job much quicker with a reference from my previous employer. It also gave me a timeframe to deal with. I usually found another position before my last day of employment. Usually, it was upper management individuals who saw me as a liability and seemed to find a reason to convince me to leave. They didn't want me there anymore and I could tell by the way they behaved around me. It is very uncomfortable to work in an environment like this and stress is a deterrent for me at any job. It wasn't that my job performance was poor, especially in the retail world. I have some impressive customer service awards.

Most frequently, management individuals recognize that my diabetes can create emergency situations. Such situations are viewed as a liability and I am asked to quit or resign. This has happened at least twice in my lifetime. They have more pressing responsibilities than my diabetes and quality customer service is implied in retail. It all depends upon the manager. The ones who had people skills were promoted very quickly. I usually retained a reserved attitude to avoid unnecessary responsibilities and hypoglycemia. My motto is: If you can't find something about your job that you really like, then you shouldn't be there

I keep hoping that I will win the lottery one day because I've had so many problems trying to fit into normal work environments with very little success. It seems impossible to win the lottery but someone is winning. I believe in miracles because I am one. Interviews for another JOB make me so nervous that I can't function. Retail businesses come and go, and my benefits diminish consequently with other employees just like me. I've found that I work best alone. That is why I decided to write this book because there is no one telling me how I should do it. Once I complete it, that could change. Everything I am telling you about in this book comes from me and my experiences, good and bad. The diabetes has had a very negative effect on my life. I've tried to do what I was supposed to do, most of the time.

Some things are completely out of my control, like my metabolism and my hormones. It seems like even work stability is out of my hands. If stress did not interfere with my metabolism so dramatically, sometimes, it would be much easier for me to function in a working environment. Unfortunately, stress is a part of work and life. But with the diabetes, anything can happen to me at any time, without me even knowing it. I do the best that I can and have done very well over the last five years in a working environment. Hopefully, the next five years will be just as productive. I was hoping to be working in a pharmacy somewhere with my current employer by now. Our store was supposed to upgrade by June 2009 but has been postponed due to our failing economy. Until our economy improves, I will stick with my part-time job and look forward to what the future holds for our business market.

Type two diabetics have an advantage in this part of life over type one diabetics. Most type two's have already established their professional lives before they develop the disease and it's ramifications on their work schedule and environment. They usually have already established a position where there is more leniency in scheduling. As a type one, you've been a diabetic since you were a child and before you were old enough to work. Establishing yourself in a professional environment is difficult when you eventually become unconscious from hypoglycemia and it diminishes your chances of acquiring that vital promotion. I am not saying that type two's don't experience such episodes. They do but later in life, after they are already settled in their professional lives.

When you are a child, parents and adults often try to influence your decision about your professional life. They often try to give advice about this and most of it for me was related to my diabetes. It had nothing to do with my skill set. Finally, the decision was left up to me. I guess no matter what form diabetes comes in, it will have an affect on the working environment and the working environment will have an affect on it. Scheduling was my first priority and almost every job I considered had a poor schedule. Then, I stopped worrying about the schedule and started looking more closely at the job requirements. Neither method has reaped the rewards I expected. Now that I have been trained to work in a pharmacy and can't find a job due to hiring freezes and our deteriorating economy, I feel that the system has failed me, yet again. Some things are out of my hands.

I am a workaholic, once I get involved in my work and the people I work with. As a retail employee, I try to observe sales that will be important when I am cashiering to help my customers and our store. I do plan to work in a pharmacy in the future when our economy improves. I just can't be sure when this will happen. While I am waiting, I can share what I know with you and hope that a light bulb will go off above your head when something happens, to you or someone you know with diabetes, and it suddenly makes sense. That is how a lot of situations became evident to me. Particularly with hypoglycemia.

Many of the symptoms look like things that you experience normally every day. For example, yawning, fatigue, laughing, sweating, blurred vision and feeling dizzy which sometimes comes from not having my glasses on. I can't predict everything and hypoglycemia looks like a normal bodily function when it's actually hypoglycemia. It takes years of practice to know which is which. There are so many distractions in a working environment that my symptoms can distort my diagnosis.

My attitude can change very quickly, especially when my diabetic needs present themselves. I try to keep them quiet but this isn't always possible. My personal statistics show that it will rear it's ugly head once every ten years, in a working environment. I have approximately five more years to go. I am trying to concentrate on my work but my body has given my mind a warning, of sorts. So, I am trying to work and concentrate on my diabetes at the same time. I keep a vigilant eye on my diabetes and things at work seem to operate better. Keeping myself under control isn't always easy when my diabetes tries to tell me something. It's the times I don't hear it that get me into trouble.

Chapter 23: Scheduling

There are many assets to working but the schedule is the most important to me and the least obliging to the employee. If I didn't have diabetes, I wouldn't care so much about the schedule. Scheduling creates the most problems for me while working. Even though my schedule may have me working from 11:00A -4:00P and they are open from 8:00A – 10:00P, my diabetic eating schedule will not change. This means that lunch should be around 12:00 noon for me and dinner should be around 6:00 in the evening, regardless. My body will tell me when I need to eat. What I dislike most about hourly jobs is the schedule.

I might be scheduled to work from 11A – 4P and then we get busy at 4:00 and they need me to remain during that influx. Retail work demands this commitment from it's employees. I usually need to stick to the hours I am scheduled to work to control my diabetes. It is never intentional when hypoglycemic shock occurs. I am not saying that this will happen every time I go over my hours. Mental notes of my past eating schedule makes it possible for me to work overtime. My diabetes dictates what will happen.

If my blood sugar is not running low around that 4:00 time period and I feel ok to work, then I probably will. If it looks like the lines will lessen without my help pretty quickly, then I may not volunteer to stay. Anytime around 12:00 noon or 6:00 at dinner can be questionable for me. My body is on it's own time schedule, depending on when I ate last, how much insulin I took, and how busy I've been since then. If I am not acting like myself, then something is wrong. Hypoglycemia is disguised by yawning and laughing which aren't seen as abnormal to the untrained eye. I don't recognize it myself sometimes. When I get low, I am either too jovial or too irate. OR I'm nervous and confused very easily and can't focus, no matter how hard I try. And, most often, I am just yawning and laughing. People who truly know me can tell when I'm not quite myself. My mother can attest to this.

I have found that if I work the same schedule all the time, I can plan my day around meals and my body responds in accordance. If my schedule is not the same consistently, then anything can happen. Crazy schedules or schedule changes don't work well for me. If I have arranged to work for someone at least a day in advance, then I am aware of it and plan for it. If it's a last minute thing, then I am not available to make that change. I always plan to be ready for work before I get there. These days, I am using my pockets to carry all personal things into work. This will include my keys, some money, debit card, tic tac candy, and my insulin pump occupies one of those two pockets at all times. My pockets are full while I am working. I try to be prepared for hypoglycemia at any time. Because our hours have been decreased due to a decline in store sales, I have not been scheduled for many hours in the past few weeks.

When I have been called at home to go into work, I am usually either gone or already have other plans and am therefore, unavailable because I didn't know they wanted me to come in. It is difficult for me to change my time once I've started my day. I've already eaten at a certain time and taken my insulin for it at a certain time. Therefore, I was planning to have dinner at a certain time to compensate for when I ate lunch. If I take a work schedule that interferes with when I am   supposed to eat dinner, then I need to ensure that I am prepared for it. It isn't necessarily that diabetes is complicated. It requires constant attention for me to avoid mishaps. My experience is vintage when it comes to the maintenance of my diabetes. It is my goal to retain ancient experiences and upgrade, when appropriate.

Banker's hours would be ideal for my condition. I would already have eaten breakfast before getting to work and only need to eat lunch at work. I would be home by dinnertime. The hours at a bank aren't usually subject to change and don't fluctuate much. I tried it for awhile through a temporary agency in a call center. My hours were set when I agreed to take the assignment. Until the assignment ended. I loved that job because I always knew when I was going to be working. My hours never changed and no one ever called me to come in.

If they were going to need me on an occasional Saturday, they would ask me about it on the Monday before that Saturday. Scheduling is always a concern of mine and not because I want it to be. I start planning my day the moment my eyes open from sleeping every day. My day always revolves around diet, insulin, and exercise. I understand the science behind these three things and can adjust each, as needed, within reason. Minor adjustments occur frequently and daily.

I was at work, one day, and I had a HYPERglycemic episode. I was thirsty all afternoon nd constantly going to the restroom. My throat and mouth were parched and my lips were chapped. This was a job where I was working face-to-face with my customers at a cash register. It was a day where I worked a ten hour shift. When I got off, I went to the bathroom first. Then, I bought a diet soda and drank it in the break room after clocking out from work for the day. From there, I went straight home which was fifteen minutes away.

I passed a McDonald's on my way home and went through the Drive-thru for another diet soda. I was convinced by then that something was wrong. When I got home, I was running a 541 mg/dL. I had changed my infusion set before going into work that day. The cannula had been inserted it into an area of subcutaneous skin tissue in my stomach that wouldn't absorb any insulin. I had gone ten hours without any insulin and my body was telling me so. My mind just couldn't recognize it until I got to McDonald's which was within five minutes from my house.

When I got home, I immediately changed my infusion set again in that same day and was feeling better within an hour later. Usually, I change my infusion set once every five days. Hyperglycemia happens less frequently than hypoglycemia. If I am certain that I am in a hyperglycemic state, I can prevent it from escalating by taking insulin. Then, it will depend upon how quickly my body can absorb the insulin into my bloodstream. The insulin will decrease my blood sugar level and then, I can regain control of it. When hyperglycemia occurs, I need to attend to it immediately and not ten hours later. Otherwise, my blood glucose will continue to rise and I could be hospitalized for no less than three days. Hospitals are expensive and I don't have the time or money to be in one, as a patient. Ignoring either condition is not an alternative for me.

Hyperglycemia and hypoglycemia can make a dent in any schedule. They are unplanned and can create issues for myself and management. However, they are infrequent and unintentional. Almost like when someone slips on the floor or sprains their finger at work. Someone might call out sick the next day from work. A sprain requires a splint and some bandaging to keep the joint from swelling or being damaged permanently. Each of these situations are different and need to be handled accordingly. Any of them might require a few hours to days off from work for recovery.

What if your car breaks down? Then, you take it into a shop for a new battery, which was what they found to be the problem. Upon further inspection, they find that you need new brakes and there is a hole in one of your hoses. You only went in to get one problem fixed and they found another one. Then, you are told that they will need to keep your car for a few more days. How are you supposed to get to work without your car? Your problems seem to be piling up. Management doesn't like to deal with scheduling issues but they happen almost every day.

I saw a commercial on TV where a young man described his work day in an office. He was a diabetic also and said that sometimes he forgot to eat. Between trying to meet deadlines with all the paperwork he does in the office and all of his other responsibilities, it is easy for him to forget to eat. Like anyone who does or doesn't have diabetes, work is very demanding of your time. He was talking about a glucometer for testing blood sugar and trying to promote it. Me, I would eat first and test second. That would be the order of how I would treat my diabetes at that time if I was as busy as he is in this commercial.

In an office environment, you may have issues with co-workers not being amiable toward blood glucose testing taking place at the desk next to them. I had gone from taking insulin injections in the employee bathroom to using an insulin pump when I was working in the call center environment. That meant that I was attached to a computer and phone for nine out of the ten hours that I worked on this particular assignment.

My doctor wanted me to test more often to see how I was adjusting to using an insulin pump. I was unable to do this according to his specifications and, eventually, discovered an alternate plan of action. It was not his recommended choice of treatment but I was able to appease him. I am still using the insulin pump even now and appreciate it's convenience and accessibility. The insulin is always there for whenever and whatever I eat and need. The biggest issue is knowing how much insulin to take for what I am eating. I am in total control of my diabetes with the insulin pump, otherwise. The insulin pump makes my diabetes much easier to manage. I can handle the diabetes. Work schematics falls under management control.

I think that working and being able to work is a wonderful thing. A person feels productive and gets more from work than money. Relationships are developed that may extend past working hours. Friendships formed often include social gatherings like parties and church events. These types of relationships can last for a lifetime. Full time work, especially, provides additional benefits like vacation time off with pay. There is also paid sick time off when needed. If you need a paid day off to help you to get rid of a bad cold, then you can utilize one of your sick days off with pay.

Work benefits often include a retirement plan and medical insurance which every family needs. Working provides us with more than just money. Anyone who doesn't realize this hasn't seen the opportunities that working can offer. However, making enough money from where you work can be the deciding factor between staying and leaving a job.

Money was never high on my list of priorities, until I started getting older. Unfortunately, the diabetes remains to take priority over any work-related issue that I may have, even money. I've found that I need more money than I did when I was a teen-ager. Bills can increase but salaries have a tendency not to when you need it. So, money is important but so are the relationships established there. The biggest issue I have with work is it's need to dominate my time for monetary, insurance, and retirement benefits. With all of the responsibilities that I have at home, some of those hours spent working could have been spent staying at home so that the refrigerator or washing machine could be repaired. Or the kids could be taken to the doctor. Then there's yard work, vehicle maintenance, home repairs, and a multitude of other things that need to be addressed at home. I do believe that working is wonderful, especially in establishing lifetime relationships.

However, I think working at home can be just as important sometimes, even if it doesn't pay anything. Part time work has been ideal for me, except for the pay, because of the diabetes. Work schedules never seem to be flexible enough to accommodate my diabetic needs. My needs can occur at any time and I've lost jobs because of my needs. There doesn't seem to be a perfect answer to this scenario in my life but I keep trying. I am hoping that I will find one of two possible careers soon. One is sitting right here telling you my stories and the other is working in a pharmacy. The pharmacy has a schedule too but it isn't as rigorous as a retail schedule.

Scheduling should be taken seriously when you consider a job. Too much availability can create stress. Learning to control stress has been an epiphany and a liberator to me. I would be able to deal with my diabetes when I need to without getting frustrated over employer protocol. Scheduling can present problems for me and I attempt to communicate clearly with the person responsible for scheduling. Retail hours are extensive and can fluctuate dramatically with store sales. Hopefully, scheduling issues can ensue as a part of my past with my future career changes in writing and pharmacy. My mission of having control over my own schedule will have finally come to an end.

Down Time with Diabetes

Chapter 24: Dreams

There is nothing normal about the kind of dreams that I had when I was a child. I don't usually remember my dreams because I sleep soundly. It's when I first awaken and am not fully awake that I remember the most. Starting at age nine, I used to have dreams about what my life would be like without diabetes. As a child, when you dream about not having diabetes, all you think about is the food that you cannot eat. With diabetes, cakes and cookies are cut out of your diet almost completely. You can eat the sugar-free cookies but they don't taste quite the same. You are taught to eat fruits, like apples, for sweets instead. From the time I was nine until I was sixteen, I ate, drank, and slept diabetes. Everything else revolved around it. It was an all-consuming part of my life.

When I had a real sweet tooth, I liked plain M&Ms. Chocolate was my favorite food and I dreamed of chocolate cake and those chocolate popsicles that my friends and brother liked. Also, those Charms lollipops with the bubble gum in the middle. I liked the Cherry and Grape flavor the most. With gum, I've always liked Trident, mostly, in the Original flavor. I tried Juicy Fruit and didn't like it. It was extremely too sweet for me and I stuck with Trident and still do today. I liked Cokes, for a short period of time, and switched to Diet Coke which I liked even better. I abandoned it and developed a taste for TAB soft drinks, in the pink can with the little yellow stars on them. The bottles of TAB were tall with little yellow stars on them too. With the diabetes, I abandoned so many foods that I had liked and dreamed about them constantly for about a year or so after I was diagnosed.

In my dreams, I could see how wonderful those foods were and remembered some of the foods I'd loved before my diagnosis. After about a year, the dreams became fewer of the foods I couldn't have and started including more of the ones that I could like the fruits that I really liked, such as apples and grapes. I liked Grape lollipops, right? I was taught, as a diabetic, to eat my main course and decline desserts. And I still eat this way, even today. The only time I indulge in a dessert-type food is when I am hypoglycemic or having cravings.

I also had dreams about learning how to give myself an injection to make my parents proud of me. I was so afraid to do this and I knew it would help my dad, especially, if I could find the courage to do it myself. I felt bad or inadequate because I couldn't do it. But I also knew subconsciously that my dad and I would grow apart after I learned. I had these dreams for two years until I started giving my own shots. Then, those dreams stopped. My dad was excluded from this part of my life and I began dealing with my diabetes alone.

Once I got past the dreams about food and my injections, my dreams of dealing with my friends came into focus. I used to dream of birthday parties that I would have gone to if I weren't a diabetic. The kids in my class would talk about the birthday parties that they went to over the weekend and what they did. Food was always included in these conversations. Suzy had the chocolate cake with the Cinderella on top of it and lots of ice cream in different flavors. I could only have the vanilla flavor of ice cream and those conversations led my mind to something that I wanted and couldn't have. These conversations spilled over into my dreams. Did I mention that I don't have normal dreams?

Anyway, for children, lots of candy and treats would be a normal dream. I am sure that some of the kids in my class didn't have a clue about my dietary restrictions, and others did. During this time in my life, this is when I found out who my real friends were. They didn't try to make me feel bad by talking about things I couldn't have in my presence. And if we did talk about it, we were more open and honest with each other. My real friends wanted to learn more about diabetes so that it wasn't foreign to them and we could learn about it together. My friends who didn't want to talk about it and didn't seem to care became distant and were placed on my Do Not Call list. This was the worst list one could ever be placed on as a pre- teen-ager.

By the time I was twelve, my food dreams weren't as frequent. My older brother was allowed to have sweets but we didn't have as many in the house because of me, from his perspective. As I moved into my middle school years, my dreams took on a different motif. They revolved around school activities and the people in my classes. I still had to make sure that I ate when I needed to and had learned ways to avoid becoming hypoglycemic. I was involved in church activities and library events more, as well as physical education stuff. I was a very busy twelve year old.

I would choose to enter an event because of the people involved in it and not because it was an event that I was good at. I wanted to be in the events or clubs that included my friends and the popular kids, like all pre-teens. When a teacher had a group of us together, we could choose who we wanted to do a project with and I wasn't usually chosen first. Even when there was someone who was my friend in the group, they often chose someone else to be their partner. I partnered with whomever was left over.

My dreams started including the different club activities and I was always excluded from the clubs organized by the popular kids. So, I chose other activities instead. I did a lot of tutoring and helped other classmates to read. I wasn't great at it but I was good at determining why they were having problems and could pass the information along to their teacher. I worked with their teacher to help them, which was when I started getting the reputation of being the teacher's pet. I had a few dreams about this too but I knew that I was helping these middle school aged children to read, even if it did tarnish my reputation. Most kids didn't like me anyway so it really didn't matter. I was actually smarter than most of them and I think they were actually jealous. But being a diabetic, I always wanted to be like one of them without diabetes which would make me ordinary like them.

I was so involved in after-school clubs that my parents had a hard time keeping up with me. I did too but that didn't stop me from getting involved in more activities. I wanted to weed out the ones I didn't care for and stick with the ones that I excelled in. It was kind of hard, at first, because there was a boy that I liked in the Science Club and I didn't surpass others in that club. I stuck with it for awhile and eventually, dropped that club. It became more work for me and I didn't enjoy it anymore, even if he was a member. I had dreams about that club often and had reservations about giving it up. The things that I worked the hardest for usually reaped the greatest rewards. This was the Science Club and I found it very interesting, besides the fact that I liked Matthew.

He wasn't extremely good-looking or a football quarterback. He did have blond hair and blue eyes which was appealing to me, at such a young age. He was nice to me and didn't treat me any different than he did his own friends. That was what I liked most about him. We had been paired together to do a project and I was determined to see it through. I wasn't a quitter. But after our project was over, I dropped out of the club. I decided that I would pursue it when I felt that I was ready for it.

My dream about Matthew and my project involved us coming up with an invention or a cure for diabetes, from my perspective, and Matthew was interested in something different. He liked airplanes. He wanted to invent a high wind velocity contraption with wings. When we actually got together to talk about our inventions, this was what we decided to do. The dream was a virtual reality. He really wanted to design his own airplane. I had a dream about us working on his invention together. We were out in this huge, grassy field with his newest flying airplane, of sorts. We were under a big tree with a picnic basket. Matthew had brought his older brother, Isaiah, to help him fly his plane.

On the first dry run, the plane didn't get off the ground. Isaiah tried it again and the plane looked like it wanted to take off but remained grounded, even on the second dry run. Isaiah was a little heavier than Matthew by ten pounds so Matthew decided that he wanted to try to fly it for himself. He couldn't get it off the ground either. This was when he decided to change the wing shape to get the result that he desired.

In reality, this was his idea and I was letting him supervise his idea and the results. I had very little knowledge of airplanes and felt that he would be more capable and successful with his invention than myself. I made a few minor suggestions but he was the mastermind of his own invention. My invention was more medically oriented and would involve more work than his did. I needed to do more research to find a feasible and believable cure for diabetes. Matthew didn't think it could be done but I presented my ideas anyway when the deadline for our demonstration arrived.

The teacher who was the director of this club liked my idea and Matthew's. We didn't win any awards for what we did but, for me, I had thoroughly enjoyed being Matthew's partner in this assignment. This was a dream that I had when I was in middle school and was starting to get interested in boys. I found that dreams didn't have to be ALL bad. By middle school, my life was becoming more normal and so were my dreams. Sleeping became more sound and deep for me.

As an adult, I still have dreams that have never been fulfilled. At this time in my life, it's possible that they will remain forever only in my dreams. My divorce crushed the biggest dream of my entire life. I don't hold any grudges toward men because of what happened between my ex-husband and myself anymore, as I did for the first three years after my divorce. I did promise myself that things will be different the next time around and that was thirteen years ago now. Life does move on and you know that things can only get better. Even my dreams are better than they once were. Dreams can become reality but remember to keep the special ones in your heart, mind, and dreams. If my dreams were any indication of my future, then love remains to have a place there and I have many more things to look forward to.

Chapter 25: My Favorite Things

The starred items below are my absolute favorite things to do. I began practicing methods of reducing stress when my doctors suggested it to my parents and myself. If you know some of my favorite things, you can see how certain situations might be more stressful to me. My favorite things are not stressful and this brings a sense of peace into my life. In situations where I need feelings of peacefulness, being aware of some of my favorites is good to know. It depends on the situation. Interviews and work circumstances are two of my most apprehensive situations. A part of being human is forming habits that fit into your life, whether good or bad. At any rate, we all have our likes and dislikes and these are some of mine. The more stars you see, the more I enjoy it.

****Sleeping Late****- For me, this is around 10:00AM. This habit started when I was eighteen years old and diagnosed with hypothyroidism. Six months before I was diagnosed, I would sleep until almost noon or later which was unusual for me. I was working and going to high school at the same time. Usually, I made sure that I got eight hours of sleep every night but not more. Due to everything that I had been taught about diabetes, reducing stress was one of the regiments toward better control of the disease. Getting enough sleep was a part of stress reduction. Once I was diagnosed and treated for hypothyroidism, I started getting into more normal sleeping patterns. I don't mind sleeping late, sometimes, because I know it's good for me. Anything that can help to support a healthy immune system and lifestyle, I will support and participate in. Also, I always feel good after a full night's sleep. I tend to be kind of a night owl and the only way I can get eight hours of sleep is to get up a little later in the morning. It is quieter at night and I like the solitude.

Watching Television- Night time television is better than daytime television for me, no question about it. After 12:00 noon, TV seems to be geared toward the soap opera and talk show stuff which bores me to tears. I think it's because I don't believe half of what is being discussed. I like mostly medical shows where diagnostic formats dominate the show plots. For example, Dr. House and CSI, Miami. Sometimes, I will watch the Discovery channel, depending on what their subject matter contains. I can watch the same medical shows over again because I recognize a lot of the medical terminology. I have been taking pharmacy classes and medical terminology was one of the required classes. I enjoy watching television most frequently after dinner. My day is winding down and my mind is leisurely deciding what I need to and will be doing for the next day.

Attending School- I've been going to school most of my life, including continuing education after graduating college. When you attend school, you are exposing your mind to different subjects and expanding your knowledge of whatever you choose. Some classes require a lot of teamwork and others are more individualized where only you are responsible for your own work. I favor the individualized ones more but the teamwork ones allow you to interact with your classmates and are usually fun. Reading is a quiet pastime done late at night to avoid that phone call that interrupts the best part of the book. Reading can open doors in your mind and your imagination can bring that story to life. It can remind you of a time not long ago and open a door that you thought was closed forever. School encourages such activities which will stay with you all your life.

****Walking****- My favorite type of exercise is walking. I need exercise in my life because of the diabetes. Exercise helps your body to metabolize or burn calories and uses less insulin to accomplish this in the meantime. I am one of those people who will park my car as far away from the store as possible. I park my car at a store on the end of the mall and walk the entire mall, just to walk. Walking isn't the most aggressive form of exercise that I can partake in but it does help to burn calories and reduce blood sugar. The best form of walking is probably with the aid of a treadmill. I have opted for walking around a track and will be using one in the very near future.

*****Shopping*****- I hate to admit it but I really do like to shop, especially window shopping. It gets me out of the house and I can shop without feeling like I have to buy something. Sometimes, I do it for the exercise. I feel like I am doing something while I am walking around the mall. There are many times when I shop because I need to. Errands are a big part of this and mine involves going to the grocery store, bank, pharmacy, cleaners, library, good will, and ETC. All of these errands are an opportunity for me to get out and walk. I take it every chance I get.

Driving- I learned at the age of eighteen how much of a privilege it is to drive. Once upon a time, especially when I was in my teens, I used to experience seizures. They kept me from driving from the time I was eighteen until I was twenty-four years old. I believe that they were due to low blood sugar episodes but I am sure that my doctor would have said otherwise. He has recently pass away. Therefore, I love to drive. Recently, I bought a 2007 Toyota RAV (small SUV) and traded my 1996 Toyota Camry for it. Now, driving is more of a pleasure because my new vehicle has a lot of great features. The stereo is the best.

My sister sells used cars and says that it has a lot of horsepower. I don't know about that but the stereo system rocks.

Reading- has been a pastime of mine since the fourth grade. It is relaxing and I can take my time doing it. As I began writing seriously, reading became intense for me. I started noticing words used and how the book was designed. The stories in books keep my mind guessing about what will happen next. Lately, I've been reading books about family situations and how husbands have met their wives. Then, there are sections involving children, pets, and girlfriends from the past. It sounds like men have some of the same questions about entering an unending marital relationship as women. I don't believe that there is any ONE answer to making a marriage perfect. There is a lot of compromising that goes on between a man and a woman in this type of relationship. I think that truly understanding each other would make a marriage as perfect as it could be. I've read many books, collegiate and romance alike, and enjoy books about medicine and family relationships the most.

Coffee - is my first food choice for reducing stress when driving. Coffee is usually warm and soothing to me. Also, the fact that there is a coffee house on every corner and it's affordable makes it an appealing choice for such purposes. As time moves forward, it's easier to manage stress when necessary and choose a method of stress reduction that I know will work.

Chapter 26: Miracles

I am a living, breathing testimony of a miracle; a miracle created by medical science. It's hard to put into words how and what this means, exactly, but I know I've lived with diabetes longer than was expected. That horrid day when I entered the hospital in 1968 with diabetes, as close to death as I've ever been in my life is proof that miracles do happen. I have done the best I've been able to do with the many trials and tribulations that have happened in my life. After forty years of living with type one-diabetes, where every scrape on my skin or irritated ingrown hair in my skin could cause an infection and cause me to die if left untreated has been attended to somehow successfully.

Diabetes is a full-time job every day of my life. There are others who have survived with it longer than myself and I have to admire them. There are many others who couldn't survive past five years with it. At the time of my diagnosis, it was expected that I could live to be about thirty years old. I am now fifty and appreciate every day. After I started giving my own injections at age eleven, I dealt with it almost exclusively all by myself from then on. There were times when I needed hospitalization and those times I needed outside help but the daily grind of it all was my baby.

I am a miracle and didn't really believe it until about three years ago. I've dealt with it every day like that's the way life is supposed to be. But for most people, their life isn't like this. There are days when you don't want to get out of bed because you don't feel good but you aren't exactly sure why. It wasn't because I wanted to take a day off from work. I really didn't feel good. I would test my blood sugar, when I had test strips because I had been able to afford to buy them that week, and find that I was a little low. But I wasn't low enough to be feeling like I was. So, what was the culprit behind that feeling? It could have been something that I ate or the beginning of an infection.

Experience is where you learn to determine why you feel a certain way. It is easier if you have test strips to be certain that it is or is not a low but they don't come cheap. And my insulin is more expensive than my strips. Things have been rough for me before, especially economically and financially. I deal with them the best that I am able, the same as it is for many other people. We all have things that are more expensive than others and medical insurance and some of my prescriptions are included in the expensive category.

The one thing that you have to be extremely careful of and never let get away from you are infections. You have heard of diabetics losing limbs, like fingers and toes, before and infections are the root of this type of misfortune. This has never happened to me yet but there is always the possibility that it could. It is hard to deal with diabetes when there isn't enough money to afford the supplies you need to take care of yourself. I was there during my divorce, especially, and found that I couldn't deal with the stresses from that situation alone. I needed someone dependable and my immediate family was and still is my most dependable source of support.

They know what can happen to me when I forget or neglect to eat. Work has been the one thing that has created the most chaos in my life when it comes to dealing with my diabetes. Because I am duty bound to work a schedule while I am there, god help me if I ever pass out while on the job. That job is gone and herego, the money from it. I wish the government would pay me for taking care of my diabetes like I should, for twenty-four hours a day, seven days a week. It's a job with no beginning or end. They pay nurses for doing this kind of work so why not me? I don't have a degree for it but I shouldn't need one, after forty –one years of experience. I have my eyesight, all of my fingers and toes, and my kidneys are functioning. Sort of like when two people live together for ten years, it's called a Common Law Marriage.

Taking care of your diabetes should be your main priority every day. The consequences are debilitating. I am here to tell you that you need to control it so that something unforeseen doesn't happen to complicate your life even more than the diabetes. I haven't been perfect about the way I eat or don't eat or in the way I have taken my medicine or done my blood glucose tests. What I have done is watch out for those infections or anything that may or may not cause them. Even if the infection seems insignificant, it can get out of hand quickly.

As a miracle of diabetic medicine, I am here to remind you that when you take care of yourself, you are taking care of those around you. Every single day is a challenge with diabetes, but it doesn't have to be so challenging that it becomes pointless to treat it. I've felt that way at times with my blood glucose readings. They were higher than anticipated at times and I couldn't understand why. Don't let it deter you. Keep testing and see if it continues to be high. If it does, go to the doctor.

Challenges are a part of life, once you are diagnosed with diabetes. Meeting them and treating them is the best course of action. Once you have determined your course of treatment, administer it and wait to see the results. If you are newly diagnosed, be sure to call your doctor and describe your results.

The results are wonderful, when they are anticipated from your treatment and they're correct. It takes practice and with practice comes knowledge. Don't forget what you've learned because you will need to remember it when the same symptoms happen to you again. I was with a friend of mine when this happened to her.

She and I had gone on a very long walk, one that she was not used to, and she became hypoglycemic. I asked her how she felt and she said that she was very shaky and weak. She looked frightened. We were on a trail and out in a wooded area. We came upon this huge rock and she sat down to make her more comfortable. She said that she had never felt this shaky before. Her vision was blurred too. I listened for the symptoms that I was aware of and those two fit hypoglycemia, considering what she had done before she felt that way. I had a beat up tootsie roll with me in my pocket and she had some lifesavers CANDY with her in her pocket. We had carried them for me, thinking that this might happen to me. She decided to eat her lifesavers while we sat on that rock to give her time to recover from it. She was fine in about ten minutes and we left there to get her something to eat.

I knew that she would be fine after she ate those lifesavers with my past hypoglycemic experiences. She was scared because she wasn't as sure as I was that it would only last a few minutes. She is a type two diabetic and I am a type one diabetic. She controls her diabetes by taking a pill every day and I control mine with insulin, every day. They are different forms of diabetes because type two diabetes affects the pituitary gland and type one diabetes directly affects the pancreas and most people can recognize which is which by the way it is treated.

My body responds to insulin, as a type one, where hers responds to pill, diet, and exercise treatments, as a type two. This is the one part about diabetes that is the most difficult for non-diabetics to understand. Experiencing it is the easiest method of understanding it. However, diabetes isn't something that I would have wished on my ex-husband and we haven't spoken since the divorce. There is no better way to describe it. There is nothing like feeling as if you are going to die and you know there is nothing you can do yourself to stop it. I remember how this felt when I was first diagnosed with diabetes....not good. I didn't understand it well enough to know what to do for myself to feel well again. I had to learn all of this from doctors, nurses, and dieticians.

But it is truly a phenomenon that I am still here. If there had been another plan for me, He could have taken me during one of my two most critical episodes, when I was hyperglycemic and as close to death as I've ever been. But He didn't. I do believe there is a God or I wouldn't be here. There is a reason why I remain to be alive. My faith has been centered on medicine and medical technology for a long time. God must have created the people who made medical technology improvements available and therefore, for me. God exists. Only He knows why I am here. Maybe my book is the reason.

All I've known for the past year when I started on this book was that I wanted to share some of my good and bad experiences with you. This way, you could hear it from someone who has diabetes and not a doctor who has never experienced diabetes and it's symptoms themselves. Some of them possibly have but none of my doctors in forty-one years were diabetics. They talk about what you should and shouldn't do, only knowing what they studied and were taught about in school from a book. Most likely, they are not physically aware of how it can truly affect them. God has bestowed a true exaltation on anyone who doesn't have diabetes, even my doctor and ex-husband.

Learning about diabetes and experiencing it gives you two different perspectives of the disease and experience, according to what I've heard all my life, is the best teacher. Diabetes is challenging, exhausting, unusual, interesting, and requires continuous attention from the moment that you are diagnosed with it. Just remember that you are here for a reason and can't be sure what it could be. You could be here to teach others about the illness or to become an astronaut but you have to discover that ultimate purpose for yourself. If you put your heart, mind, and soul into something that you truly believe in, that objective may present itself with little effort on your part.

I believe in the medical miracles of my life and I am here to say that they are magnificent, and come in all sizes, big or small. Walking, reading, sleeping late, drinking coffee, shopping, they all seem like normal every day activities. Who knew they could all be so good for me. However, they all help to control my disease and I enjoy them. I discovered such things thirty years ago and still practice them today. Coping with diabetes doesn't always have to be complicated. Efforts made to establish such practices can create a lifetime of healthy habits. I believe in extraordinary occurrences and I'm here to say that they are well worth any effort expended, barring death, to accomplish them and are magnificent.

Chapter 27: Giving Up is Option-less

This chapter is hypothetical and controversial and gives the mind other things to consider about diabetes. These are beliefs, not necessarily factual, that I have formed after living with diabetes for so long. I can't give up on my diabetes, as others I've known before me. I have been struggling with it for forty-one years and probably will for another forty-one years. I know there has to be a conclusion to the suffering of diabetes out there somewhere. Someone is coveting it. Often times, money is the primary reason why something as important as a cure for diabetes is not publicized. Or maybe there is too much to be lost by making it public including money. Number one, It could put endocrinologist physicians out of business and Number Two, whoever has it may feel that they can get more money for it by waiting for the most exorbitant offer.

I found a website online that states that alternative medical therapies have been suppressed to the public for such diseases as aids and cancer, and other chronic life threatening sicknesses. (1, 2) What if type 1 diabetes is one of them? This could be considered speculation to some people but what if it were true? Why is it that our pharmaceuticals have increased by thirty percent, or more, over the last ten years? I do believe that money is the ruling factor in such medical matters that could possibly save someone's life. There can be many different reasons why the treatment hasn't been made available. I've seen information on healing type two diabetes. There is a lot of literature to support this concept for type two but not type one. There are at least 24 million people that have diabetes in the US, according to another online article that I found. (3) Many diabetics could be healed and never have to suffer with it again.

Some doctors would cease to exist without diabetic patients. They could change their specialty and switch or move into some other area of medical practice. If you are a diabetic, you do not have the same option. The diabetic patient cannot make such a switch to become non-diabetic, most frequently. Type 2, maybe, but not type 1. Insurance procedures monitor patient treatments and is controlled through coverage of services. Most patients go to doctors covered by their insurance. If the doctor or service is not covered by their insurance, then the patient looks for a doctor who is covered. Once discovered, your need for that doctor's prescription for insulin would diminish for insulin-dependent diabetics. Some of my beliefs are speculation that I heard from someone else somewhere along my forty-one year journey with diabetes. Research for the conclusion to this relentless disease has continued since I was diagnosed in 1968. These speculative ideas that I once heard are all theories and considered to be a part of research which is not factual until it can be undeniably confirmed. If you want to do more research, check out my resources on page 197.

I am tired of being a victim of diabetes and it's symptoms that cause me to be uncontrollably moody and unpredictable at times. I've tried to control these behaviors but my feelings of uncertainty or fear about what could happen make me nervous and pensive. It would be outstanding to be able to live a non-diabetic life without such reservations, after 41 years. It consumes my life frequently. If I don't take care of it, the consequences are dire. It really doesn't give me much of a choice. But I am still making the choice to control it.

Sometimes, I eat something that I shouldn't and I know it will affect my blood glucose negatively. However, I'm not a saint. I want to have some normalcy in my life occasionally. I like a piece of cake or a candy bar, once every six months or so. I am careful and ration my sugar intake so that it is in moderation, as recommended by my doctor and dietician. It would be nice to be able to eat something just because I wanted it. When I was on the injection, I was more careful about what I ate because my supply of insulin was limited and, now, I am using an insulin pump. The insulin pump allows me more freedom with my diet than I ever imagined possible. It functions as an artificial pancreas would. It isn't perfect but it is better than taking an injection, or two, every day.

My life has been more routine since I've been on the insulin pump. I will continue to use it as long as my insurance will cover the supply costs. Insurance is chronic just like my diabetes, but when it comes every month, it is one of my most costly expenses. The cost of one bottle would pay for two weeks worth of groceries. The pump is pretty expensive but to a diabetic, the insulin pump is well worth the money. Not having to worry about whether I have enough insulin to eat something that I want is remarkable. My insulin pump carries approximately 250 units of insulin and I replace the infusion set or tubing every four or five days. The cost of quality health insurance is rising at an alarming rate. However, I never thought I could go four days without worrying about taking an injection either. Consult with your doctor if you are interested in utilizing an insulin pump. He will know what treatment is best for you.

The demand for quality medical insurance is fourth on most people's list of monthly expenses because the mortgage for their house, keeping their kids in clothes and school, and keeping their car running will come before health insurance. The demand for insurance will drop as the cost increases and the coverage diminishes over time. With our current economy, pay increases are being withheld for another year or two. Coupled with the cost of groceries, gas, electricity, telephone, and other bills, insurance companies could price themselves right out of business. Medical insurance is a necessity when you need emergency care from a fatal car accident or even a bee sting or hives. Without medical insurance in such instances, your life could be compromised.

All of my other medicines, except my insulin and test strips, are affordable with insurance coverage. The insulin is a necessity for me and the test strips run a close second but if push comes to shove, the insulin wins. I can't survive longer than three days without insulin. If the cost of insurance continues to increase and my paycheck continues to stagnate due to a slow economy, I could soon be faced with monumental debt. I have no control over the increase in healthcare costs and my paycheck seems to be in the hands of the economy. Where do I have any choice or chances of making a positive change in my financial stability? Jobs are being lost every day with few or none to replace them. I don't have an answer to such questions but being aware that they exist helps to prepare me for what could come.

Most of my chapters mention how my diabetes was affected by my life or my life was affected by my diabetes. They are interchangeable and situational. There are several gaps during certain times in my life where nothing of any consequence really was affected by my diabetes. Those were the days that reflect an ordinary and happier time in my life. We all have bad days and most of my days are pretty typical. I treasure them and realize that tomorrow could bring on a whole new set of rules that I will have to adhere to.

On the days where something has happened, it remains in my memory and I do anything I can to keep it from happening again. Scheduling is important, when working, because mine can't be changed drastically. I plan my day from the minute I get out of bed by my diabetes and how I am feeling. If my work schedule changes, then I have to be able to make adjustments in my treatment. If I know that I can't, then I don't agree to any changes. If I am at work, sometimes, I don't have a choice as to whether I can stay or not. According to management, I need to stay. According to my diabetes, I can't always be sure. Not without testing my blood glucose first.

Otherwise, I don't need to worry about when I eat or take my insulin because I am not at work and not on someone else's schedule. My time belongs to me when I'm not on the job. The body functions on it's own time clock, whether I am at work or not. This is true whether you have diabetes or not. If you have diabetes, then you have to be more alert to bodily symptoms. If I am not at work, then I can attend to my diabetes when I need to. If I am at work, I don't make any unnecessary changes to my schedule for that day.

Diabetes is very demanding of my time, even moreso than work. This is why I wish the government could allow me some monetary compensation for a full-time responsibility. When I am taking care of myself, I am caring for those around me as well. If my metabolism is in overdrive and causes my hypoglycemia, then I need to detect it before I am passed out in the floor. This is my responsibility, 24 hours a day/ 7 days a week. If they know that a cure for diabetes is available, I wish they would finance it to make it available to people who need it. This would be the second most incredible event to materialize since I was diagnosed in 1968. The insulin pump was the first. I will continue to struggle with it for as long as I have to and won't give up. Giving up is not an option.

Clarification From A – W

Chapter 28: Summary of Important Points

I want to re-cap some of the more important points mentioned in some of the experiences I've had with diabetes. I've mentioned hypoglycemia and hyperglycemia frequently in my book. These two diabetic conditions control whether I will have a good or disastrous day. Hypoglycemia happens more frequently than hyperglycemia. This has been evident from day one of my diagnosis. Hypoglycemia can surprise me but most often, it doesn't. Controlling hypoglycemia is much easier than controlling hyperglycemia. Hypoglycemia can be controlled by eating a piece of candy. I will respond to the sugar very quickly. Hyperglycemia can only be controlled through the administration of insulin. And then, it depends upon how long it will take my body to absorb it before I can begin to feel like myself again. Both conditions are exhausting but hypoglycemia is much quicker to recuperate from. Hyperglycemia can last for several days.

<u>Take care of yourself and listen to your doctor</u>. Keep in mind that the methods listed below have worked best for me. Your doctor knows you better than myself. Take your doctor's advice and try different treatments recommended. Finding the one that works best for you is your job.

<u>Hypoglycemia</u>. Diabetes does have a few of it's own rules and regulations and as you live with it every day, you realize that you have to follow them. I've had diabetes for forty-one years now

and have experienced several low blood sugar episodes at the most unfortunate times. Several times, I was at work. You can't always be ready for it when it happens, especially at work. When you are not at work, your time is your own. Do what you need to do for yourself to keep your day from becoming a disaster. I become very uninhibited when I get dangerously low and can't keep my attention in focus for very long. I will, sometimes, laugh uncontrollably or am overly nervous. This is when I know I need to do something. Your symptoms may be much different than mine but you know what they are when you recognize them. Remember to consult with your doctor about any changes in your treatment.

Hyperglycemia. Hyperglycemia is much more serious than hypoglycemia. Hyperglycemia gives you that yucky feeling until you've had diabetes as long as I have. My mind doesn't recognize the frequent urination and thirst as abnormal for me because it's seen these symptoms before. I've had diabetes for so long that I don't get as many psychological warnings as I did when I was a child. Explaining diabetes to non-diabetics is challenging. There is no way for them to understand what you are going through because they can't experience it. Their body naturally makes enough insulin to compensate for what they consume. They don't need to maintain normal blood sugar levels every day and won't experience the effects of hyperglycemia. Remember what your symptoms were before you became hyperglycemic and contact your doctor as soon as possible.

Preparation. Diabetes can make you feel like a victim because it can obscure your day very quickly. This may happen only one day a year for you and all of your other days are fairly routine. Because of how debilitating it can be on that one day a year, it keeps me aware that it is always around. I can't lose sight of this and if I do, the diabetes will remind me. Therefore, I am a victim of it's effect on my body and mind. This is why I learned how to prevent these difficult days within the first two years after my diagnosis. If I feel like hypoglycemia or hyperglycemia is a possibility, then I ensure that neither happens. It's my responsibility to myself and others around me to do this. I am very careful in my daily treatment of my diabetes so that I know, beforehand, if I am too high or too low. The problem with the way I think is that I think that I am prepared for it all the time. Then something unforeseen happens that I couldn't have been prepared for and the diabetes shows me otherwise. However, it doesn't happen to me as frequently as it did when I was first diagnosed. What I've learned about it over the years has helped me to quickly diagnose what I need to do.

Recalling what happened and why is something you should talk with your doctor about. Sometimes, all you need to do is call.

Test Often. I keep a constant check on my blood glucose to see how I am doing. When I am at work, this isn't possible because there is no place for me to keep any personal belongings. My monitor and test strips are temperature-sensitive and can't be in excessive heat or cold for more than six hours. Therefore, I go by how I feel when I am at work. Most of the time, I can tell when I feel low but this isn't an accurate assessment of my situation. BG monitoring is the best way I know how to keep myself from getting into trouble. I realize that the test strips for the glucose monitor can be expensive. Use them sparingly and keep a close check on your blood sugar levels. Knowing this can make your day much more pleasant than the few nightmares I've had with hypoglycemia. Good habits are hard to come by but frequent BG testing is, without a doubt, one of them.

Driving. The few times I felt uncomfortable driving myself to work, I caught a ride with a friend. By the time I got to work, I felt fine. That has only happened to me twice in forty-one years. But when it does happen, see if someone can give you a ride. The consequences can be detrimental and embarrassing, in the long run. You could even lose your driver's license and that could cost you your job. I try to be as careful as I can, when it comes to vehicular situations. Use your best judgment and decide what's at stake for you. I always did and my instincts haven't proven me wrong yet, even after forty-one years as a type one diabetic.

Avoid Stress. Remember to keep a close eye on stress. Different situations may affect you in a way that you are unaware of. It may not affect your life in the way that it has mine but it is still a good idea to watch your stress levels. They can play havoc with your blood glucose level and you don't want to end up passed out in the driver's seat from hypoglycemia. Some diabetics experience high blood glucose levels from stress. My blood glucose plummets downward when I am stressed. I found that when I was hypoglycemic behind the wheel, it was far too difficult for me to even determine how I was going to get off the road. The last thing I could focus on was what I was going to do before my mind could recognize that something was really wrong. It should have been the scariest twenty minutes I've ever experienced, if I'd been aware of it. But I was unaware of what was happening outside of my car. My objective was to get where I was going and no other thoughts could be formulated except for my destination. Confer with your doctor about any stress-related incidences.

Work schedules. If you are a diabetic and have an hourly schedule, these hours are usually subject to change. With diabetes, your bodily schedule is not subject to change. Your body is still going to get hungry at breakfast, lunch, and dinner like it is supposed to. That is when the drops in blood glucose can become most evident to you. With practice, others will not notice it. I try to explain this to whoever is doing scheduling and co-workers nearby. I am usually labeled as a problem when it comes to scheduling and am being cautious and considerate of others. Some jobs have stationary hours upon your date of hire, not subject to change. This is your best case scenario. With time at work, you can determine which schedule is best for you.

Expectations. As I mentioned before, life has never been the same since I was diagnosed. I keep living life the best I can every day. The longer I am here, the better it is for other diabetics, especially where insurance issues are concerned. We've gotten a bad reputation from somewhere and I intend to stay around long enough to get involved in something to help improve this situation. If I live to be eighty years old as a type one diabetic, I believe I will have long outlived my life expectancy upon my diagnosis. The way I see this is that life has too much to offer and I'd like to see it improve for diabetics. Life is good, whether I have insurance or diabetes or not, and I am looking forward to any future improvements.

Being a diabetic is a 24/7 job. You should be able to recognize hypoglycemic and hyperglycemic symptoms when you see them. I can't mention these two conditions enough to emphasize their importance on your well-being. If you don't see it coming, then you've fallen down on the Security part of your job. I fell down on that job a few times myself mentioned in Jailbait, Ah, Belligerence, and Control. I had other times that I mentioned but I caught them before they became critical. Take care of your diabetes and yourself. You will have many more good days than bad, once you can identify their signs and symptoms and foresee them coming. If you are overwhelmed with hypoglycemic episodes, contact your doctor.

X, Y, Z -   These are factors that are specific to you that may affect your diabetes. X may be that you are a male where I am affected by the Y factor, as a female. Another example is that stress causes my blood sugars to drop and other diabetics are affected oppositely where theirs increases. Find out what they are and understand them. They could be the key to keeping your diabetes under control. Control is what it takes to have more normal days than not. I like my normal days and so does my doctor. Do whatever the doctor says to determine your X, Y, and Z factors. Particularly during that first year after your diagnosis.

Remember that diabetes **Can** be controlled. I didn't really believe this was possible during my first twenty years with diabetes. The harder I tried to control it with recommended strategies, the more impossible it seemed to be to accomplish. It is easier to foresee control with the improvements made in the field of medical technology. I went to see my diabetic doctor last month. She informed me that if more diabetics were doing as well as myself, she wouldn't have a job. This made me feel good and made me realize that control really is within my reach. Just remember that it is possible and don't give up trying to find the right regiments for yourself.

Choose a good doctor. He or she can make your expenses affordable and life easier to live. Be sure to take a proactive approach to your treatment. Know what it is that you want to talk about before you go in to see your doctor. What methods work best for the doctor don't necessarily work the best for you. Medicine is costly and so is everything else we need. Work with your doctor and he or she will reciprocate the same efforts. If you ever need treatment outside their expertise, they can give you a reference to another doctor for extended care. This will be one of the most important relationships you will ever establish.

I am a diabetic, not a doctor. I am not viewing diabetes completely from a medical standpoint. I am someone with years of experience in living with diabetes. Living with it is complicated and explaining it is even more frustrating. There are some very good points to remember mentioned, if you are a diabetic. However, certain situations may affect you much differently than they have me and how you treat them is up to you and your doctor. I have dealt with the daily regiments of diabetes on my own because no one else could see how it affected me, unless I was passed out somewhere. I decided very young that I wasn't going to allow myself to become hypoglycemic to make a point to someone else and counteract it's effects before they became serious. If they cared or were curious enough about my condition, they would check into it themselves to find out more. Remember to take care of yourself, FIRST.

Final Note

I've been a diabetic for forty-one years and it's hard being a diabetic in a non-diabetic world. I have a lot of experience with diabetes and felt that it was time for me to share some of them with you. School and work responsibilities are exigent when diabetic conditions present themselves. Other environments are easier to control because you aren't accountable to them for your time, like church. If you need to leave, you can. At work, your time is theirs when you are scheduled to be there. Diabetes demands your time and attention no matter where you are. If you are newly diagnosed, time will be your teacher. Learn well and take careful notes. Most of those notes will remain in your memory.

I am not a doctor but my experience is priceless. It can help you to understand why things happen the way they do sometimes. I am taking you through certain times in my life and this is where gaps in time are evident. I wanted to focus primarily on critical diabetic situations that happened in my life, or close calls. If you are familiar with diabetes, my book will be more enjoyable for you. Questions specific to your condition should always be directed toward your personal physician. They can help you to make decisions for yourself about your treatment. This will be one of the most important relationships you will ever establish. Take care of yourself and talk to your doctor. Your life depends on it.

Reference Source(s):

(1) http:// www. webring.com/ hub?ring = medcure, p.79

(2) http:// www. theogian.net, p.79

(3) http:// findarticles.com/p/articles/mi_qn4188/is_20080625/ ai_n27518086/

ATLANTA -- The number of Americans with diabetes has grown to about 24 million people, or roughly 8 percent of the U.S. population, the government said Tuesday.

About the Author

Britta McGeorge has been a diabetic for forty-one years. She is known as an eccentric intellectual. Diabetes has contributed to her eccentricity. She became interested in writing while in college with Public Speaking as her most recent accreditation. Writing is her passion developed through time created from successful diabetes management.

www.ingramcontent.com/pod-product-compliance
Lightning Source LLC
Chambersburg PA
CBHW070534200326
41519CB00013B/3040